Management
of Procurement

Management
of Procurement

Edited by
Denise Bower

Published by Thomas Telford Limited, 40 Marsh Wall, London E14 9TP
www.thomastelford.com

Distributors for Thomas Telford books are
USA: Publishers Storage and Shipping Corp., 46 Development Road, Fitchburg, MA 01420
Australia: DA Books and Journals, 648 Whitehorse Road, Mitcham 3132, Victoria

First published 2003
This paperback edition 2010

Also available from Thomas Telford Books
Financing infrastructure projects. T. Merna and C. Njiru. ISBN 07277 3040 1

www.icevirtuallibrary.com

A catalogue record for this book is available from the British Library

ISBN: 978-0-7277-3221-7

Typeset by Helius, Brighton and Rochester

Preface

Procurement is the process of acquiring new services or products and includes contract strategy, contract documentation and contractor selection. It extends to all members of the supply chain, including those responsible for operation and maintenance. The Association for Project Management (APM) *Body of Knowledge* states that 'The procurement strategy should include potential sources of supply, terms and types of contract/procurement (for example, partnering or alliancing – versus commodity purchasing), conditions of contract, the type of pricing, and method of supplier selection.'

The need to be able to prepare and develop a procurement strategy is essential for any project management professional and an appreciation that this must be done very early in the project life cycle is paramount. This can only be done effectively when those involved understand all of the options available and the implications of the choices they make. Hence, this book covers all aspects of procurement, from drafting and using contracts to procurement strategies for privately financed projects. It is written from a construction perspective with examples from other industries used where appropriate.

For many projects, procured goods and services form the highest percentage of expenditure and so it is important to achieve value for money through careful appraisal and management. Relating to how well the project scope is defined, the state of the supplier market and the perceived level of risk associated with the project the procurement strategy will promote a controlled and auditable response to external influences and ensure that the project, objectives are met.

The purpose of the editor and authors is to provide information about the various areas that comprise the procurement process to support those professionals tasked with undertaking this job in the real world. This book does not detail the legal issues relating to contracts, although a very brief summary of the main points of law is included, recognising that there are a large number of law books already available. It does offer guidance, explanations and case studies to help the reader comprehend the strengths and weaknesses and realistic meanings and outcomes of the stages in the development of an effective procurement strategy.

This book brings together the theory and practice that relate to procurement and offers guidance on how the process should be undertaken. This should ensure that the project manager has clear information on which future decisions can be based. This guidance should prove to be a useful point of reference for those developing a procurement strategy or project execution plan to ensure that all viable options have been considered and the strengths and weaknesses of the chosen route are understood. The book begins by outlining the role of procurement in the construction industry, including the problems it faces and some of the steps that can be taken to overcome these problems. It goes on to consider contractor selection and raises issues relating to the allocation of risk in contracts and factors to consider when awarding a contract. A brief summary of the relevant legal aspects is also provided.

Chapter 3 is written by a lawyer and describes the drafting and use of construction contracts from a legal perspective. The role of the 'wish list' is discussed and guidance is provided on how this can be successfully managed. Project constraints and the management of risk are also considered. The book then goes on to examine how a client can procure the services of a project manager, covering such issues as terms of appointment, pre-qualification, selection and award. Contract strategy is explored in detail in Chapter 5 and all of the major organisational arrangements and payment types are explored. A summary of the key features of the most common standard forms of contract is also provided. This leads into a chapter on the incentivisation of construction contracts, an increasingly popular way of motivating contractors. The relationship between benchmarking, key performance indicators and incentives is explained and a framework for developing incentive mechanisms is provided.

Collaborative forms of working such as partnering, alliances and joint ventures are explained in Chapters 7, 8 and 9. The chapter on alliancing takes the form of a case study to demonstrate in detail how such a strategy can be implemented. Moving towards more innovative procurement techniques, private finance procurement routes are explained in Chapter 10, and then this is followed by an overview of project value systems and strategies that can be used to achieve them. Procurement through programme management is addressed in the penultimate chapter as consideration is given to the strategic nature of procurement. The book finishes with an assessment of future trends in construction procurement, proposing that supply chain brokers will be the key to future success.

This book is unique in that it provides a holistic view of procurement from the perspectives of both practitioners and academics. It should promote a better understanding of this complex area by providing detailed descriptions of the process to allow project managers to implement successful projects.

Denise Bower

Acknowledgements

I am particularly grateful to my co-authors for all of their efforts in the production of this book. I would like to thank particularly Professor Nigel Smith of the University of Leeds for his advice and guidance on the form and content of the book, which has led to it being both comprehensive and easily accessible. I would also like to thank Professor Stephen Wearne of the Centre for Research in the Management of Projects at UMIST for his personal contribution and for co-directing the EPPE module on Contracting and Procurement over the years at UMIST, upon which much of the material is based.

I would like to acknowledge the pioneering work into contract strategy and incentive mechanisms undertaken at UMIST by Professor Peter Thompson, now retired.

Finally, I would like to acknowledge the contribution of the many students that I have supervised over the years, including Fotis Skountzos, Chryssa Sarri and George Hagan. Their work in the area of partnering has been a great help and has made the writing of this book much easier.

List of contributors

Denise Bower, BEng, PhD, MASCE, ILTM, is Professor of Engineering in Project Management in the School of Civil Engineering at the University of Leeds. She was formerly the Shell Lecturer in Project Management at UMIST. She is a leading member of Engineering Management Partnership, which offers diploma and master's degree-level qualifications to engineers of all disciplines. She was a member of Latham Working Group 12 and has an extensive record of consultancy work with clients in construction, process engineering and manufacturing.

Edward Davies, LLB(Hons), MSc, is a partner based in Masons' Manchester office, specialising in process, energy and infrastructure projects. He is a visiting research fellow at UMIST and a mediator trained by the American Arbitration Association and accredited by CEDR. He is a joint editor of *Dispute Resolution and Conflict Management in Construction – An International Review*, published by E&F Spon.

Paul Garthwaite, MEng(Hons), read civil engineering at the University of Leeds, and obtained his masters degree in 2003. Having a number of years' industry experience in the rail and highway sectors, he has worked under a number of collaborative and contractual agreements throughout his career. His recent research has included the manipulation of framework agreements within multi-project environments and the influential elemental aspects of partner cooperation, partner complementariness, trust building and their effects on ownership balance and structure.

Martin Graham MSc, PhD, MCIOB, is currently responsible for a £120 million NHS Trust development programme to be deliverd over a period of 8 years. His responsibilities include converting the Trust's long-term strategies for improved performance into a series of deliverable projects. The requirements are to extend the estate and improve working practices. Previously, he designed and delivered similar programmes in the university sector and in the hotel, food manufacture and water supply industries, and spent many years delivering construction projects.

Chris Hallam, LLB(Hons), is a solicitor based in Masons' Manchester office, specialising in non-contentious work for the construction, engineering and infrastructure industries, including PPP project work. He contributed a chapter to *EC and UK Competition Law and Practice: a Practical Guide*, published by Sweet & Maxwell, has had articles published in *Legal Week*, *BTI Journal* and *Sales Director Magazine*, and regularly writes for the Masons website.

Benjamin Joyce, MEng, graduated from the School of Civil Engineering at the University of Leeds in 2002. His final year dissertation was on incentive mechanisms in construction contracts in Europe. He has taken time off to travel the world, and has returned to work for a major civil engineering consultant with a view to furthering his interest in project management.

Steven Male, BSc, MSc, PhD, holds the Balfour Beatty Chair in Building Engineering and Construction Management, School of Civil Engineering, University of Leeds. His research and teaching interests include strategic management in construction, supply chain management, value management and value engineering. He has led research projects under the EPSRC IMI programme 'Construction as a Manufacturing Process', with the DETR, with the DTI and within the European Union 4th and 5th Frameworks. He is a visiting professor in the Department of Civil Engineering, University of Chile. He works closely with industry and has undertaken a range of research, training and consultancy studies with construction corporations, construction consultancy firms, and blue chip and government clients.

Nigel J. Smith, BSc, MSc, PhD, CEng, FICE, MAPM, is the Professor of Construction Project Management in the School of Civil Engineering, University of Leeds. After graduating from the University of Birmingham, he has spent 15 years in the industry, working mainly on transportation infrastructure projects. His academic research interests include risk management and procurement of projects using private finance. He has published 15 books and numerous refereed papers. He is currently Dean of the Faculty of Engineering.

Phil Spring, BSc(Hons), MAPM, MCIArb, is the managing director of Spring and Company Limited. He was chairman of the North West Branch of the Association for Project Management from 1999 to 2001, a council member of the Association for Project Management from 1999 to 2001 and chairman of the North West Project Management Forum from 2001 to 2002.

Stephen Wearne, BSc, PhD, CEng, FAPM, PMP, was formerly the project engineer and project manager of large projects in South America and Japan. He is now a visiting senior research fellow at the Centre for Research in the Management of Projects, University of Manchester Institute of Science & Technology, and an emeritus professor at the University of Bradford. He is the founder chairman of the Contracts & Procurement Specific Interest Group, UK Association for Project Management, the author of books and papers on project organisation and control, emergency projects, contracts and joint ventures, and the editor of the APM guide *Contract Strategy for Successful Project Management.*

George White, BSc, CEng, MICE, MIStructE, graduated from UMIST in 1970, after which he worked for Sir Alexander Gibb & Partners and George Wimpey. where he was involved in and led dock and harbour and power station projects. In 1976 he joined Conoco and held supervisory and managerial positions on four large North Sea projects. One of the projects used an alliance contract strategy that he developed and was instrumental in executing. In 1997 he moved to Halliburton Brown and Root, where he held a senior managerial position and continued the development of the alliance contract strategy and conceived a set of management processes for executing projects on a total life cycle basis. A later position as a global capital projects manager at Blue Cement saw him both improve global project delivery and implement a company-wide performance improvement initiative. In 2002 he joined BG-Group as a project manager working on mega-projects. Throughout his career he has lectured and published major technical papers.

David Wright, MA, CIChemE, ACIArb, left Oxford with a degree in jurisprudence and spent 30 years in industry. He gained experience in the automotive industry, the electronic industry, the defence industry and the chemical engineering and process industry. He was the commercial manager of Polibur Engineering Ltd and, in the mechanical engineering sector, was European legal manager to the Mather & Platt Group. He is now a consultant on matters of contract and commercial law David is also a visiting lecturer at UMIST and a visiting fellow in European business law at Cranfield University.

Contents

CHAPTER ONE

The role of procurement in the construction industry

D. Bower

Introduction

Recent years have seen high levels of turbulence; companies that were market leaders a decade ago have in many cases encountered severe reversals of fortune. Rapid advances and complexity in technology, and the accompanying growing uncertainty in the business environment have brought about mergers and takeovers, and these have changed the shape of many markets. Traditional barriers between industries are breaking down. Inevitably, this has given rise to a very high level of competition and complexity. There is also a growing demand from the marketplace for ever-higher levels of service and quality.

In response to this changing business environment, there has been a search for an instrument that would offer a sustainable competitive advantage. In other words, companies are now seeking a position of superiority over competitors in terms of customer preference. The emphasis in business has swung towards business strategies that have the creation of long-term customer loyalty as their central focus. Business leaders are pursuing new business paradigms that allow their companies to work closely with their traditional and new business partners in order to adapt to the rapidly changing marketplace.

These new business relationships are arrived at through developing procurement strategies that balance work, motivation and risk for long-term, sustainable performance improvements. Procurement is the process of acquiring new services or products and includes contract strategy, contract documentation and contractor selection. It extends to all members of the supply chain, including those responsible for operation and maintenance. This chapter describes the construction industry, its key features and the problems that it faces. A brief overview of the role of contracts in projects and the organisation of traditional relationships is provided. This is detailed in later chapters. The industry's response to the problems it is facing is also outlined, and then these issues are explored in greater depth later in the book.

The construction industry

The construction industry plays an important role in the economic development of any nation. Construction in the UK, for example, is one of the pillars of the domestic economy. According to the Construction Task Force (1998), the industry in its widest sense was likely to generate an output of roughly 10% of the country's annual gross domestic product (GDP) in 1998. In addition, it employs about 6.4% of the total labour force.

The construction industry is a large and highly diverse sector of industrial activity, ranging from the construction of large power plants, through the construction of large residential and non-residential buildings, to the small-scale renovation or repair of existing facilities. In the UK, construction is the largest industrial sector, about three times larger than agriculture, with construction companies comprising 50% of the companies registered. A major part of the UK construction industry works overseas and generates millions of pounds of overseas earnings.

The construction industry encompasses different types of work. The variability of the industry's works is reflected in a comprehensive definition provided by the UK's Department of the Environment (1995):

> Operations such as building, civil engineering and specialist contracting – including bricklaying, plumbing, heating and ventilation contracting, electrical work, carpentry and joinery, painting, roofing, plastering, glazing, scaffolding, specialist work in suspended ceilings, floor and wall tiling, insulating, and reinforced concrete work, as well as other activities where the major element of work is building, civil engineering, or the installation of products and systems, either in buildings or in association with civil engineering works.

The construction industry may be seen not as a single industry but as consisting of several different market areas. There may, in theory, be a uniform knowledge of the construction market and perhaps what competitors are doing, but in practice the industry may be seen as a series of overlapping markets defining a particular service divided by size, type, geography and complexity of work. For example, between the small repair and maintenance builder and the large contractor, there is virtually no competition; they have their defined market areas. These may be based not only on the type and size of work but also upon the location, and both the small builder and the large contractor might have to make strategic choices, which may confine them to one or two market areas.

Falling trade barriers have prompted construction companies to operate outside their traditional boundaries; for example, expatriate construction firms in developing countries are often in classes of their own. In general, a large number of relatively small enterprises carrying out work on a local basis make up the construction sector, while a very few large engineering and contracting firms are of national and multinational character. The UK construction industry contains almost 200 000 contracting firms, of which about half are private individuals or one-person firms. Only 10 000 contracting firms employ more than seven people. Current employment levels show 1.46 million employees, roughly one in ten of the UK working population.

Historically, construction has been project-oriented in its organisation and management. Construction project tasks are frequently one-off undertakings and custom-built to specification. Each project marks the establishment of a new and temporary production system and organisation located at the point of consumption. In other words, there is comparatively little continuity in the form of production system established between projects in the construction industry. With this single one-off project characteristic, the question then arises as to what extent construction companies can be expected to demonstrate improvements. The process of production in the industry, to a large extent, is inseparable from the geographical location of the output that is produced. That is, the finished product cannot generally be transported, since it is produced at the point of consumption. Thus, the production process itself is almost always at the mercy of physical environmental conditions.

The organisation and management of a construction project almost invariably involve interlinkages between a number of organisations involved to varying degrees and in varying ways throughout the total project 'cycle'. The temporal nature of the construction multi-organisation means that each complete project has to go through a series of contracting and procurement processes, and there is a tendency towards the use of varied companies for projects procured by a single client. Typically, different procurement and contracting relationships can be found between the client and the main contractor, the client and the specialist consultants, the main contractor and the subcontractors, the main contractor and the specialist contractor, the main contractor and suppliers, the suppliers and the manufacturers, and so on. Subcontracting is a widespread practice in the industry; there is high level of subcontracting by main contractors to a large number of small, specialised firms.

A typical construction project passes through a number of stages, from inception through to completion and commissioning. The construction stage is only one part of the overall project. Thus, in managing the construction works, it should be recognised that there are networks of internal and external relationships.

Contracts are the basis of managing engineering projects. Therefore, the type of contract strategy chosen must take into account the project objectives and the characteristics of the parties to the contract; and aim for an equal distribution of risks and responsibilities. The main aim of a contract is to clearly outline the risks associated with the project and how they will be allocated for the project life and not just for the design and construction phases. The client defines the contract strategy, where objectives and the roles of the project team members are considered. The primary goal of a contract strategy should be to achieve the client's objectives. This can be accomplished by incorporating:

- client involvement
- allowing for changes
- motivation of contractors
- best risk allocation
- cash flow of the client and contractors.

Careful planning is extremely important during the process of contract selection. The client must give careful thought to what influences its decision to enlist a contractor's services for a project. Then these reasons must be categorised in order of importance. Smith (2000) lists some factors used by clients when about to retain contractors for a project. The client aims to take advantage of the contractors' skills and to get contractors to carry some project risks. The most suitable contract strategy for a project must be established based upon a structured, in-depth and efficient analysis of all relevant factors (i.e. all available options). Issues that affect the selection of a contract strategy are:

- clearly defined project objectives from the client
- responsibilities of the parties to the contract, which must be accurately stated
- risk allocation between the parties involved in the contract
- payment mechanism
- incentive mechanism to secure a proficient performance from the contractor
- motivation for the client to supply the necessary data and support to the contractor

- client having enough flexibility to add changes
- clients being able to methodically assess change in a fair manner.

To measure the amount of work, motive and risk transfer needed to assist in choosing the appropriate contract strategy, Smith (1999) advocates charting all existing options, as shown in Figures 1.1 and 1.2. A variety of organisational structures are available; in practice, some organisational structures are closely linked with a particular type of contract, for example, the traditional approach with the admeasurement contract. As this is not always the case, it is preferable to consider the decision on organisational structure as separate from, but interrelated with, the decision on the type of contract.

Every time an interface is introduced to the project organisation the management effort required to deliver a successful project is increased, as is the risk of failure. The aim should be to minimise the number of interfaces between the different organisations. The organisational structure must define communication and contractual links. Barnes (1983) surmises that there should be principles of risk allocation in contracts that would reduce the number of disputes between contractors and clients, since project objectives would be better achieved. He went on to propose that risk and incentives are directly proportional, so in order to maintain a contractor's motivation to perform, some risks should be transferred to the contractor. Hence the risks the contractor bears should be enough to maintain an incentive, but not so much that it is unfavourable to the contractor or the client.

Features of the construction industry

There are a number of features that separate the construction industry from other industries:

- construction involves a high volume of specialist work and a wide range of trades and activities
- many of its projects are often one-off designs and lack any available prototype model
- construction lacks repetition and standardisation of designs and components
- the arrangement of the industry is such that design has been separate from construction traditionally
- the industry is highly labour intensive
- the product is often manufactured on the client's premises with the work being subject to interference from the physical environment.

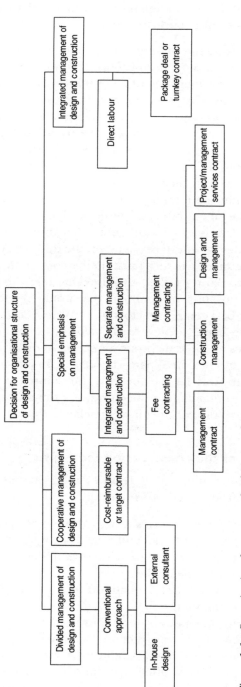

Figure 1.1. Organisational structures for design and construction

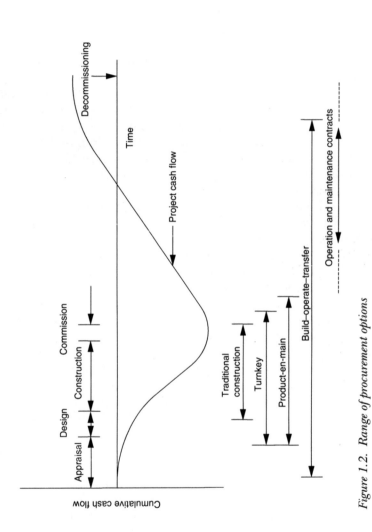

Figure 1.2. Range of procurement options

A unique feature of the industry is the use of standard methods of contracting and tendering serviced by teams of independent consultants, civil and other engineers, architects and property and quantity surveyors, who all provide an agency service between the clients and the contractors. Construction suppliers plug in at various levels of the construction process. The interactions between these independent bodies on a project have been described as complex, a peculiar characteristic of construction.

Once the outline design and the budget are approved by the client, the design team is ready to carry out a detail design and prepare detailed drawings and specifications. The design and specifications often go through several changes and modifications. The team also prepares tender documents that set out what the contractor is required to do, with penalties imposed for failing to deliver. A number of contractors and, in turn, their suppliers are invited to bid for the construction works against the drawings and specifications, traditionally on the basis of a bill of quantities. This is often prepared by a quantity surveyor and it contains fairly complete specifications of the required work, as well as a schedule of rates for the works. Thus construction companies often prepare tender bids from inadequately detailed information. In most construction projects, this is the first time – the tendering phase – that the contractor is brought into the project.

In many traditional cases, especially in public works, the lowest complying bidder ultimately becomes the successful builder selected for the works. There is a tacit understanding that this situation prevails because of 'public' accountability. Knowing that selection is on a lowest-price basis and that changes during construction are inevitable, contractors and suppliers initially bid below cost to win the contract, but then they find it easy to raise the price because of the changes to specifications during the works. Once a builder's bid price is accepted, then that company is formally contracted to undertake the construction work as main contractor.

The successful construction company then plans out the construction works, which involves selecting a site team, establishing a detailed works programme and resource schedule, placing orders for materials and equipment, and arranging subcontracts with specialist companies. The planning is usually carried out by an office-based team of planners, buyers, surveyors and engineers. There is very little involvement of site workers in the planning process, and that is so even at a later stage when works have actually begun on site. However, on site, it is often the site workers who take full responsibility for the management of the works in accordance with the project plan, liasing with head office staff and members of the design team as deemed appropriate.

Problems facing the construction industry

The industry's problems can be regarded as falling within three categories: the demand issues, such as low and discontinuous demand; the supply issues, such as inefficient methods of construction; and some common issues, such as poor management and adversarial culture. However the industry's problems have been examined, existing literature tends to agree that, compared with other industrial sectors, the construction industry has proved to produce low and unreliable profitability through low performance in terms of high costs, overlong project timescales, poor durability and high life costs. Recent surveys have shown that clients have generally been dissatisfied with construction outputs and these problems are peculiar to and inherent in the nature of the industry. Some of these problems are now described in detail below.

Inadequate investment in training, research and development

Too often in the past, the construction industry has been singled out for criticism of the training and development of its workforce. A number of research studies have reported findings from the UK construction industry that it invests very little in training, research and development. This is, however, the case in many other countries, both developed and developing. With today's rapidly changing business environment, there is the need for any industrial sector that wants to remain competitive to change its business structure and improve the skills of its workforce in order to meet the ever-higher, growing and sophisticated demand from customers. Large construction firms in Japan have given a relatively high priority to research and development activity; this has been attributed to the fact that they often compete on the basis of distinct technological solutions to construction problems and not on strategies to cut down costs. This contrasts with other countries, whose focus has been on cutting down costs to the minimum level possible, often by pushing risks on to other parties irrespective of their capabilities to manage them.

The temporal nature of construction employment, which is often project related and demands mobility, presents difficulties for the typical small and medium construction enterprises in developing workers, considering the high cost of training. In addition, the fluctuating and uncertain nature of construction demand or workload might have led to contractors' reluctance to employ large numbers of workers on a more permanent basis and maintain high levels of training and staff development. The outcome has been a cheap, unskilled, temporary labour force, leading to poor quality and less value for money. Not only would a commitment to training and research into 'best practice' be of immense benefit to individual workers and their firms, but also the standards in the industry as a whole could be raised.

Fragmentation

It has widely been reported that extreme fragmentation is a particular trait of the construction industry. As much as fragmentation is seen in the number and size of construction firms, it may also be observed in the diversity of professions and trades in construction. In other words, a number of distinct disciplines are required to work together in order to complete a single construction project. In the UK, for example, Rafferty (1991) reported that construction companies constitute about 51% (180 000) of the 350 000 registered companies. Latham (1994) quoted a figure of 200 000 contractor firms in 1992.

The various parties who provide input to construction are often represented by a number of separate bodies. In the UK, for example, a body such as the Building Employers Confederation represents main contractors. Professionals such as engineers and architects are represented respectively by bodies such as the Institution of Civil Engineers and the Royal Institution of British Architects. There are several other bodies that represent specialists such as mechanical and electrical engineers and trade contractors. Similar bodies also exist in the US. The Associated General Contractors of America and the National Association of HomeBuilders are two examples. More often than not, there are conflicting interests and this inevitably affects the construction process and therefore construction efficiency. There is no doubt that poor management strategy and inappropriate contracting and procurement routes in this fragmented industry structure have been a major contributor to adversarialism in construction relations. This, in no small way, has contributed to the poor perception of the industry. In contrast, the Japanese construction industry is characterised by a network of tiered, interlocking supply relationships (known as *keiretsus*).

The demand side of the construction industry is, in general, less fragmented. In the UK, the establishments of the Construction Clients Forum and the Construction Round Table are perhaps providing the demand side with more or less a unified interest. But in general, construction clients are not a homogeneous group and their objectives therefore vary.

The construction industry is also characterised by functional divisions. There is a split of responsibility for design and construction. This is a natural constraint to efficiency and innovation. Unlike the producers of goods in other industries, the contractor traditionally has no input to the design of the facility that has to be constructed.

Adversarial relationships

For many years, the construction industry has had a poor reputation in relation to its adversarial nature of relationships. The Chartered Institute of Purchasing and Supply (CIPS) has argued that the adversarial problem is prevalent at all levels in the construction organisation. The UK construction industry spends as much as £3.3 billion yearly on disputes. The Construction Best Practice Programme has specifically stated that there is evidence that, in recent years, relationships between clients and contractors in the industry have become increasingly adversarial. They recognised this as the cause for the increasing number of disputes and growth in litigation in the UK. This is a diversion of management attention from constructive collaborative teamwork into the management of disputes and litigation. This is certainly accompanied by reduced productivity and increased costs.

Inadequate involvement of suppliers

The relationship with suppliers is a crucial aspect of the construction business. It has been estimated that between 70 and 80% of the value added in construction projects is contributed by subcontractors and suppliers below the top tiers of the construction 'supply chain' (Warwick Manufacturing Group, 1999). This means that the products and services provided by the main contractors' suppliers typically account for up to 80% of the total cost of the construction project. Any failure to perform by a supplier has a direct and potentially serious impact on construction efficiency and profitability, and therefore the way in which products and services are procured and in which their delivery is managed has a significant impact on the outcome of the construction project. Recent studies and experience have shown that suppliers' involvement in design and cost development is a major opportunity for cost savings and quality improvements.

Large number of small and medium-sized enterprises (SMEs)

The construction industry in many economies is polarised with a small number of large firms and a vast number of small to medium-sized firms. Although the large construction companies undertake the major works, the smaller firms undertake a substantial proportion of the total workload. In Japan, for example, although the industry tends to be dominated by five major contractors, existing data show that the top ten firms account for only 13% of the total market, while the top hundred contractors have only a 30% share of the market. A similar situation has been observed in the US and the UK. In the UK, research has shown that almost 95% by number of construction firms have less than seven employees, yet they undertake over 30% by value of the national workload.

Major firms are known to prefer engaging several specialist subcontractors, rather than employing direct labour for the execution of construction activities. It has been estimated that the number of suppliers on a contractor's database could be between 500 and 2000 for a building, and much more for a general civil engineering construction. This makes it very difficult to raise standards throughout the supply chain and to manage the supply base effectively. There are several reasons for this choice, which are primarily rooted in the nature of demand in the industry. A number of research findings have expressed concern about the problem of changing and unpredictable demand for construction. Construction firms find it difficult to fully utilise their workforce when demand is low and falling. They are forced to use all kinds of strategies to compete for scarce and discontinuous workloads. In view of this scarcity, clients take advantage and become keen at buying at the lowest price and contractors then find their margins eroded. Consequently, these construction firms shed their staff and resources through redundancy and outsourcing to reduce their oncost, and become more dependent on a network of subcontractors, whom they may manipulate by pushing unfair risk burdens onto. As a result, capabilities diminish and construction products are of inferior quality. When the demand cycle turns to the high side, the supply capability is found to be lacking and firms attempt to meet their labour requirements through more and more subcontracting. Thus, there is an infiltration of unskilled and inexperienced suppliers establishing themselves in order to meet the high demand, and again the consequence is poor construction quality and adversarialism. In general, the structure of the construction industry is to a large extent dictated by the nature of construction demand.

The industry's response

The preceding sections have described the major issues that the construction industry is facing. It can be inferred that new construction processes and commercial relationships are required in order to improve the links between the project stages and the procurement and management of construction suppliers. The industry has in the past decade responded by taking action. Both practitioners and academics have been seeking what has been called 'best practice' in construction. In the UK, for example, the Construction Industry Board (CIB) instigated the Construction Best Practice Programme and the Movement for Innovation (m4i). They produced codes of practice and good practice guides for practitioners. The Achieving Competitiveness Through Innovation and Value Engineering (ACTIVE) engineering construction initiative has produced the ACTIVE pilot workbook, which identifies 17 construction best practices. The Research and Development Group of the Building Down Barriers Initiative has developed a new approach to construction procurement – prime contracting. The UK is not alone in pursuing such ambitious improvement goals in construction. It is known that the US construction industry has even more demanding aspirations: the Construction Industry Institute (CII) has been carrying out research to improve the competitiveness of the industry. Presently, CII has developed 11 best practices (including partnering, alignment and team-building); has pro- posed 12 best practices pending validation (including organisational work structure and employee incentives); and has written 24 information topics, including supplier relationships, contract strategies and the global construction industry. Also, in South Africa, the Public Sector Procurement Reform Initiative relating to the construction industry is one of the wider initiatives currently under way in the country. This list is not fully inclusive but it offers an insight into the industry's response to its unacceptably poor state.

Summary

This chapter has revealed that today's business environment places great challenges on industries, particularly on their responsiveness to rapidly changing market demands. Although the construction industry plays an important role in business and development, its performance in responding to changes is far below that of other industries such as the manufacturing, aerospace and automotive industries. The industry's unique nature may contribute to this low performance. However, many of the underlying causes may be related to its traditional structure and organisation. The industry's problems are caused by many factors, such as adversarial relationships; fragmentation; inadequate investment in training, research and development; inadequate involvement of suppliers; and the large number of small and medium enterprises. In many parts of the world, there have been efforts to raise the industry's image. Organisations and groups have evolved and are championing the cause of improving the performance of construction. The current contracting and procurement practices in the industry are being given critical examination in order to improve upon them. Many of these developments are described in the ensuing chapters.

Bibliography

Barnes, M. How to allocate risks in construction contracts. *International Journal of Project Management*, 1(1) (1983), 24–28.

Construction Best Practice Programme (1999). Website: http://www.cbpp.org.uk.

Construction Task Force. *Rethinking Construction*. HMSO, London, 1998.

Cox, A. and Townsend, M. *Strategic Procurement in Construction*. Thomas Telford, London, 1998.

Department of the Environment. *The State of the Construction Industry*, issue 4, July. HMSO, London, 1995.

Latham, M. *Constructing the Team: Final Report of the Government/Industry Review of Procurement and Contractual Arrangements in the UK Construction Industry*. HMSO, London, 1994.

Movement for Innovation (1999). Website: http://www.m4i.org.uk.

Rafferty, J. *Principles of Building Economics*. Blackwell, Oxford, 1991.

Smith, N. J. (ed.). *Managing risk in construction projects*. Blackwell, Oxford, 1999.

Smith, N. J. (ed.). *Engineering Project Management*, 2nd edition. Blackwell Science, Oxford, 2000.

Warwick Manufacturing Group. *Implementing Supply Chain Management in Construction*, Project Progress Report 1. Department of the Environment, Transport and the Regions, London, 1999.

CHAPTER TWO

Contractor selection, contract award and contract law in the UK

D. Bower

Introduction

A client has the ultimate responsibility for project management. The client must define the parameters of the project, provide finance, make the key decisions, and give approval and guidance. The contractor (supplier) provides a service for the client. These parties must work together if a project is to succeed but, through the proliferation of claims, clients and contractors have become further removed, the construction industry has suffered, projects cost more and clients look elsewhere to invest money.

The three main functions of contracts are work transfer (to define the work that one party will do for the other), risk transfer (to define how the risks inherent in doing the work will be allocated between the parties) and motive transfer (to implant motives in the contractor that match those of the client). There is a basic conflict between these provisions, and this chapter concentrates on the allocation of risk at contract award and outlines the key legal aspects that impact on these relationships. The drafting of contracts is explained in more detail in the next chapter.

Known and unknown risks in contracts

The identification and allocation of risk is a lengthy process that will require a number of iterations for optimum results. During project appraisal, risks that may occur throughout the whole life of the project should be identified for the whole supply chain. These could then be considered on the basis of:

- which party can best control events
- which party can best manage risks
- which party should carry the risk if it cannot be controlled
- what is the cost of transferring the risk.

That is to say, some are pure risk, for example, *force majeure*, while others are created, for example, by the technology or by the form of contract or organisational structure. These are not the same. The client must ensure that, through the contract strategy chosen, its exposure to risk is at its most equitable, considering both the up and the down side. The risk management process is dynamic to reflect the fact that risks and their effects change as the project progresses.

The impact of risk events on projects is, in the vast majority of cases, related either directly or indirectly to cost. Time delays inevitably have a consequential cost. Where materials or plant fail or the supplier of services does not perform, the additional cost is apparent. Where less tangible risk events occur, such as emissions or environmental disruption, no direct cost may be incurred immediately by the client, but in these circumstances the costs may be incurred at a later date.

Client organisations should appreciate, when deciding upon the allocation of risks, that they will pay for those risks that are the responsibility of the contractor, as well as those that are their own. Contractors employ contingencies in their tenders as a means of guaranteeing their return in the event of construction risks occurring.

The payment mechanism employed, price or cost based, will determine the location of these contingencies. Tender documents explain the allocation of the risks and responsibilities between the parties to the contract. In some cases client organisations are now requesting potential tenderers to provide a risk statement as part of their tender. The risk statement provides the client with the risks, often not covered in the contract, that the tenderer feels may occur and how that tenderer would respond to such risks should they occur.

A number of clients now list potential risks in the tender documents and request tenderers to price each of them as part of the tender. Table 2.1 lists a number of potential sources of risk. The evaluation of such risks and the price for their cover are part of the tender assessment criteria. The size of the contingencies employed by the contracting parties will be dependent upon a number of factors, which may include the following: the riskiness of the project; the extent of the contractor's exposure to risks; the ability of the contractor to manage and bear the consequences of these risks occurring; the level of contractor competition; and the client's perceptions of the risk/return trade-offs for transferring the risks to other parties.

When risk events that are the client's responsibility occur, the contractor should receive the funds necessary to overcome the particular risk event. Where there is some uncertainty over

Table 2.1. Potential construction risk sources (Smith, 1999)

Physical	Natural, ground conditions, adverse weather, physical obstructions
Construction	Availability of plant and resources, industrial relations, quality, workmanship, damage, construction period, delay, construction programme, construction techniques, milestones, failure to complete, type of construction contracts, cost of construction, commissioning, insurances, bonds, access and insolvency
Design	Incomplete design, availability of information, meeting specification and standards, changes in design during construction
Technology	New technology, provisions for change in existing technology, development costs and intellectual property rights, and need for research and development

responsibility for a particular risk event, the contractor is entitled to pursue claims for additional payment from the client upon its occurrence. Clearly the client is likely to wholly or partly pay for risk events irrespective of which party bears responsibility for them.

Contractors usually assess the cost or price of given risk events higher than clients. The reason for this is related to the long-term effects that risk events have on the business of the two organisations. This is particularly the case when a large client organisation employs small and medium-sized contractors to construct small to medium-sized projects. A cost overrun of 10% on a £1 million project would be a source of concern to a large client, but probably little more. To a contractor in the current economic climate, with low margins, it could be the difference between staying in business and liquidation.

The risk-averse behaviour of contractors and risk-neutral behaviour of major clients has been identified elsewhere. A risk-neutral client is assumed to view a £1 loss in the same light as a £1 gain. For a contractor the loss may be perceived as far higher. The effect of this is to make the contractor's estimate of the cost of a given project greater than that of the client if the responsibility for risk is evenly split between the two parties. Whilst this may not be reflected in the contractor's tender, it is likely to become apparent as the construction of the project proceeds and risk events occur.

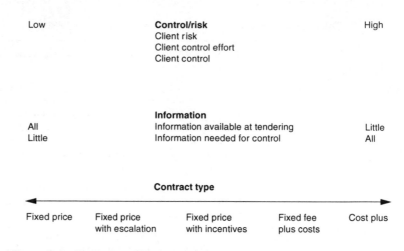

Figure 2.1. Payment mechanism options

Risk allocation strategies

There will be occasions where no contractors will be prepared to bid for a contract which places arduous conditions upon them and the client has to reconsider the contract policy. A contractor's exposure to risk must be related to the return that it can reasonably expect from a project. Thus if a contractor is making only a 5% return on a project, it is reasonable for a contractor's risk exposure to be restricted. Alternatively, tenders may be much higher than expected, reflecting the cost of transferring the risk to the contractor.

The main characteristics of the available choices of risk allocation strategy can be grouped according to organisational structure or payment mechanism, as discussed in Chapter 5 and summarised in Figure 2.1. The choice of contract and hence risk allocation strategy is determined by the policy decisions of the client and the requirements of the individual project. On occasion, however, it is the policy considerations of the client which take precedence, with little regard to the project concerned. The client must remember that inappropriate strategy on the retention or distribution of risks will jeopardise the project.

Construction risks such as ground conditions, risk of non-completion, cost overruns and risk of delay are considered as major technical risks. Most construction risks are controllable and should be borne by the contractor whether the project is publicly or privately funded. Similarly, risks associated with labour, plant, equipment and materials, technology, and management are controllable risks and should lie with the contractor(s).

Specification risk and errors in design which could have a detrimental effect on both construction and operation are also common. Physical hazards that may occur in the construction phase include *force majeure*, such as earthquake, flood, fire, landslip, pestilence and diseases.

Contractor selection

The selection of external contractors is one of the crucial decisions made by the client. The criterion for selection may be price, time or expertise. The price criterion is often the key issue, as the client seeks the most economic price for the development, whereas time and expertise criteria are often seen as being less important because of the need to expedite the construction programme and the need for good-quality workmanship. The client's objectives in the tendering process are:

- to obtain a fair price for the work, bearing in mind the general state of the construction market at the time
- to enter into an agreement with a contractor who possesses the necessary technical skill, resources and financial backing to give the client the best possible chance of the project being completed within the required time, cost and quality standards.

Before entering into the tendering process, the client must draw up a contract plan to determine the number of contracts into which the project will be divided. The basic consideration of this plan is the effect of the number of contracts on the client's management effort. More contracts will lead to more interfaces and greater management involvement, whereas fewer contracts reduce this involvement but may increase the client's risk exposure. There are certain principles which should be used when determining the number of contracts:

- the size of each contract should be manageable and controllable for the contractor
- the contract size must be within the capacity of sufficient contractors to allow competitive tendering
- the time constraints of the work and the capacity restrictions allow for the separation of contracts rather than one single contract.

The tender process may take a number of forms; the main distinguishing feature is the level of competition. Open tendering involves a high risk element for the client as many of the tendering organisations will be unknown. With selective tendering in either one or two stages, a limited number of organisations are invited to tender after some form of pre-selection or pre-qualification has taken place. In this case award to the lowest conforming tender is not such a high-risk strategy. Negotiated tendering takes place when a client approaches a single organisation, based on reputation, but this can also be time-consuming. The risk here is that at a later stage in the project the client may question whether or not value for money has been achieved in the absence of competition.

A number of factors will influence the pricing policy of the tendering organisation, such as competition, availability of resources and workload; however, these should not influence the criteria that the client uses for selection, but rather be taken into account as part of the client's evaluation. There are two stages at which the client and the project manager can control the selection of contractors (or consultants). These are before the issue of tender documents and during tender analysis, before contract award.

Both evaluations are important, but have different objectives:

- *Pre-tender* – to ensure that all contractors who bid are reputable, acceptable to the client and capable of undertaking the type of work and value of contract. Further, if an unconventional contract strategy is selected it is strongly recommended that they have experience of the approach.
- *Pre-contract* – to ensure that the contractor has fully understood the contract, that its bid is realistic and that its proposed resources are adequate (particularly in terms of construction plant and key personnel).

The most formal method of contractor evaluation is pre-qualification. A full pre-qualification procedure may include:

- either a press announcement requiring response from interested firms or direct approach to known acceptable firms
- issue by the client of brief contract descriptions, including value, duration and special requirements
- provision of information by the contractor, including affirmation of willingness to tender, details of similar work undertaken, and financial data on number and value of current contracts
- information on turnover, financial security, banking institutions and the management structure to be provided, with names and experience of key personnel
- discussions with the contractor's key personnel
- discussions with other clients who have experience of the contractor.

The evaluation may be done qualitatively, for example by a short written assessment by a member of the project manager's team. For large contracts it is preferable to try to quantify each contractor's abilities.

Some large, international clients suggest that pre-qualification should be based upon the ability to perform satisfactorily in terms of experience, past performance, capabilities (personnel, equipment and plant) and financial capacity. The pre-qualification of small, local contractors may be difficult as they may not have full financial and capability reports. These clients also suggest that the minimum period for bidding should be 90 days. This is to allow the contractor to carry out technical, labour and other investigations. Others suggest that the minimum period for bidding should be at least six months. The number of bidders should be limited as tendering is expensive. The principal terms of agreement should be known before bidding. The method statement, programme and construction plant should be identified.

A thorough pre-qualification should eliminate many of the risks that relate to the external organisation. These can be stated as:

- *Financial* – the investigation involves an assessment of the financial statements, a check on the financial exposure of the company on domestic and overseas contracts, and the history of recent financial disputes.
- *Technical* – the assessments are concerned with the current commitment of labour and plant resources, the ability to handle the type, quality and size of work at a specific time and performance on site for previous projects.
- *Managerial* – the investigation involves identifying the managerial approach to risk, contract strategy, claims and variations.

At this stage frank, open discussions should take place between all parties concerned. Risks should be identified and methods of dealing with their occurrence and contingencies to be allowed against them negotiated. A good pre-qualification should reduce the number of bidders to a maximum of six. This may be further reduced to three when genuine competition is assured.

Tender analysis and contract award

Once the pre-qualification is complete a list of tenderers can be compiled. These tenderers will then be issued with tender documentation, when this has been completed and submitted, all of the tenders can be evaluated. Evaluation is primarily concerned with the justification of the lowest-price bid that meets the client's overall requirements. It is essential that the client defines clearly and precisely the bid requirements to ensure that the submissions can be evaluated in terms of a common information base.

Commonly, tender analysis concentrates effort on the price by trying to discover high or low rates and weighting of rates. It is also essential to identify the likely causes of problems during the course of the contract, for example inadequate or inappropriate resources, an unrealistic plan or method of construction or an inadequate treatment of risk. Tender analysis should be undertaken rigorously and systematically. On major construction contracts additional time and effort at this stage can lead to much smoother contract execution. Factors to be considered include:

- That the bids are responsive to the tender documents and that any qualifications are acceptable.
- Correction of bid prices for arithmetical errors.
- Differences in technical content.
- The adequacy of resources proposed by the contractor. Clients or project managers will be greatly assisted here if they have prepared their own *realistic* estimate of time and cost.
- Differences in the timing of payments required by the client.
- The reasonableness of advance and mobilisation payments.
- The impact on project economics of any different contract durations which may be bid.
- How to assess bids containing a number of different elements, some of which may not be financially quantifiable.
- Whether to give a margin of preference to some bidders, for example indigenous contractors in a developing country, bidders who offer equipment which is compatible with existing equipment and bidders who have previously worked harmoniously with the client.
- Does the proposal meet maintenance requirements?

If discussions or negotiations are held with tendering contractors, they should be fully recorded and incorporated into the contract documents where appropriate. Before the contract is signed, the client must ensure that it is in a position to meet all its obligations.

The detailed evaluation of technical, financial and contractual information can be carried out within the relevant departments of the client's organisation, with specific checks on technical expertise, price and contractual points being sufficient to identify the three lowest complying bids. Then the client should carry out a further evaluation and, if necessary, evaluate any alternative bids.

The construction programme submitted by the contractor should be taken into account as part of the technical evaluation as it indicates the contractor's overall approach to the work although many standard forms do not require a programme to be submitted at tender. The method statements and the proposals for plant and labour levels can be submitted as non-contractual information and used to justify the bid. Often this information is used to reject infeasible bids rather than to select the optimum bidder. The client must ensure that the technical information requested from the contractor is relevant to the evaluation procedure.

While the bid price may be considered as the dominant factor, it is essential that all of the financial implications are examined. The evaluation process begins with an arithmetic check on a rate by rate basis, identifying discrepancies between each bid and relative to the client's estimated rates. In some cases one tenderer may load a particular rate based on its own calculations of the quantities required. The client may consider that some rates have been deliberately front loaded and request the tenderer to reallocate those rates without prejudice to the overall bid price. The client may also compare the estimated cash flow projections with those of the tender bids, occasionally using a discounted cash flow analysis as a further evaluation parameter.

The contractual evaluation may be carried out as a separate assessment or as a part of the technical and financial evaluation. Compliance with the contract documents is considered to be paramount. Any qualifications included in the contractor's bid which had been accepted in the initial evaluation stage would be re-examined and clarified; if necessary, the bid may have to be rejected. The contractual evaluation is summarised in a report identifying those areas of risk and possible contractual problems associated with each of the bids.

Contract law

A contract is an agreement which gives rise to obligations which will be recognised by law and enforced in the courts. (Manson, 1993)

Traditionally, contracts are classified into deeds and simple contracts. Deeds must be in writing, signed, witnessed and delivered; they do not have to include consideration to be enforceable. All other contracts, whether written, oral or by conduct, are simple contracts. The structural elements of a contract are now described.

Offer

An offer is a statement by one party with a willingness to enter into a contract on stated terms, provided that these terms are in turn, accepted by the party or parties to whom the offer is addressed. An offer may be expressed or implied from conduct and can be addressed to an individual, group or the whole world; *Carlill v. Carbolic Smoke Ball Co., CA, 1893*. An offer needs to be distinguished from an 'invitation to treat', where you are invited to make an offer. Advertisements of goods for sale are generally treated as invitations to treat (*Partridge v. Crittendon (1968)*), along with auction sales, displays of goods and the sales of land.

An invitation to tender does not normally amount to an offer to contract with the party submitting the most favourable tender although it is possible that an invitation to tender may amount to a unilateral offer if it is clearly intended to be so; *Harvela Investments Ltd v. Royal Trust Co. of Canada (CI) Ltd, HL, 1985*. An offer has no legal effect until it is accepted. Along with counter-offers and express rejection, an offer may be revoked at any time until accepted; *Routledge v. Grant (1828)*. Communication of revocation to the offeree must be made by a reliable third party if not the offeror; *Dickinson v. Dodds, CA, 1876*.

After a fixed period of time or after a reasonable length of time, where 'reasonable' is relative to the subject matter, an offer will lapse. An offer can also lapse if it is made subject to a failed condition.

Acceptance

An acceptance is an unqualified expression of assent to the terms proposed by the offeror. Introducing new terms that the offeror needs to consider is referred to as a 'counter-offer' and its effect in law is to bring an end to the original offer; *Hyde v. Wrench (1840)*. Careful differentiation is required between a counter-offer and a mere request for information. It is possible for an offeree to respond to an offer by making an enquiry as to whether the offeror would be prepared to amend some terms of the offer; *Stevenson v. McLean (1880)*.

Counter-offer analysis is also used for 'battle of the forms' scenarios where one party makes an offer on a document containing that party's standard terms of business, and the other party then 'accepts' the offer using a document detailing that other party's (conflicting) standard terms. Obviously no contract is formed, but courts have shown that if the second party's communication is accepted through conduct, for example delivery of goods, a contract has been formed through the first party accepting a counter-offer on the second party's terms; *British Road Services v. Arthur Crutchley Ltd, CA, 1968*.

Therefore a tender constitutes an offer to an invitation to treat, but acceptance does not always result in a contract. If a party invites tenders for the supply of a specific quantity of goods on a certain date, or the tender is for the supply of a specific quantity of goods over a period of time, then acceptance creates a contract. If the quantity is not specified and is to be ordered as and when required, then acceptance does not conclude a contract but is a standing offer, which is only legally accepted when an order is placed and thus a contract is then formed; *Percival Ltd v. L.C.C. Asylums Committee (1918)*.

The general rule is that an acceptance must be communicated to the offeror and this becomes valid when it is brought to the attention of the offeror through receivership. Where the means of communication is instantaneous (oral, telephone or e-mail) the contract is created when and where acceptance is received; *Brinkibon v. Stahag Stahl mbH, HL, 1983*.

It is possible to remove the need for communication in a bilateral contract for the offeror to waive, both expressly and impliedly, the need for communication of acceptance to the offeree, for example by dispatching goods in response to an offer to buy, and is thus by conduct; *Taylor v. Allen (1966)*. This is subject to the reservation that it is not open to the offeror to impose a contract on the offeree against the latter's wishes by stating that silence amounts to acceptance; *Felthouse v. Bindley (1862)*.

A final point about acceptance is about acceptance by post. Where post is the appropriate and possibly requested means of communication, acceptance occurs immediately after the letter is posted, even if the letter is delayed or destroyed; *Adams v. Lindsell (1818)*.

If the offeror states a method to communicate acceptance, with no others acceptable, no contract is formed if another method is used; *Eliason v. Henshaw (1819)*. If a communication method is stated but it is not clear that others are unacceptable, an equally advantageous method would suffice; *Tinn v. Hoffman (1873)*. Where the offer does not list a method, the appropriate method can be assumed to be the form in which the offer was made; *Quenerduaine v. Cole (1883)*.

Intention to create legal relationship

The intention to contract is a necessary independent element in the formation of a contract. Advertisements are essentially statements of opinion or 'mere puff' and are not intended to form the basis of a contract. The courts look for evidence of contractual intent and thus even a precise statement will not be binding if it is clearly not a serious statement. Domestic and social agreements have little influence on framework agreements but the main intention is to form a commercial agreement, where it is assumed the parties intend to be legally bound, unless expressly stated otherwise; *Rose & Frank Co. v. Crompton Bros. Ltd, HL, 1925*.

Collective agreements between trade unions and employers regulate pay and conditions of work. Under Section 179 (1) and (2) of the Trade Union & Labour Relations (Consolidation) Act 1992, a collective agreement is presumed not to have been intended by the parties to be a legally enforceable contract unless written expressly so.

Consideration

Along with offer and acceptance, and contractual intent, consideration is an integral part of the formation of a contract. John Lush in *Currie v. Misa (1875)* defines consideration as:

> some right, interest, profit or benefit accruing to the one party, or some forbearance, detriment, loss or responsibility given, suffered or undertaken by the other.

Past consideration is not like executory consideration where the exchange of promises takes place in the future and is considered to be invalid consideration. If the service is performed at the defendant's request and both parties agree to post-payment, the court may enforce the contract; *Kennedy v. Broun (1863)*. A person to whom a promise is made can only enforce the promise if that particular person provides consideration for that promise; *Tweedle v. Atkinson (1861)*.

Consideration must have some value and contracts are a bargain freely entered into; therefore the courts are not concerned with adequacy or fairness (except exclusion clauses). One example of this is a 'peppercorn' rent. If consideration is empty, illusionary, or concerned with human feelings, the courts feel this is insufficient, i.e. consideration must have an economic value.

Consideration is also insufficient when a promise to perform an existing duty is already owed to the promisor due to a public duty or already existing contract. Examples include *Collins v. Godefroy (1831)* and *Stilk v. Myrik (1809)*. *Williams v. Roffey (1989)* follows by allowing the consideration to be valid if existing contractual duty bestows a practical benefit on the other party. Performance of an existing contractual duty owed by the promisee is also seen as good consideration.

Terms of a contract

One of the final steps in the formation of a contract is the identification of the terms and their effect. The terms of a contract determine the extent of each party's rights and obligations, and the remedies should terms be broken.

Certainty

To create a binding contract, the parties must express their agreement in a form which is sufficiently certain for the courts to enforce;

Scammell & Nephew Ltd v. Ouston, HL, 1941. The idea is that the parties form the contract, not the courts, and in commercial contracts flexibility is necessary to keep up with market changes, so variation clauses are included, for example Clause 51 in the ICE Conditions of Contract, to provide flexibility within certainty.

Representation and terms

During negotiation of a contract, statements are classified as either a term or a representation. A representation statement induces the contract but does not form part of it, unlike a term that is a promise or undertaking included in the contract. This results in a difference whereby if a representation is broken the result will be misrepresentation, but if a term is broken it is a breach of contract.

A statement does not become a term if the statement maker asks the other party to verify its claim, whereas, if the statement is made to prevent the finding of a defect and succeeds, the courts consider this a term; *Schawel v. Reade, HL, 1913.* Following from this, a statement will be a term if the injured party would not have entered the contract had it not been made; *Bannerman v. White (1861).* Should there be an imbalance of knowledge and skill, i.e. an expert and a layman, a court may conclude that any statement made by such a party is a term; *Dick Bentley Production Ltd v. Harold Smith (Motors) Ltd, CA, 1965.*

Collateral contracts

When a contract is entered into on the faith of a statement made by one of the parties a collateral contract could be formed; *De Lassalle v. Guildford, CA, 1901.* A collateral contract can only exist if all the elements of a contract exist and can be valid even though terms conflicting with the main contract may exist; *City of Westminster Properties Ltd v. Mudd (1959).* The collateral contract is a useful tool and can be used to sidestep the parol evidence rule, to evade exclusion clauses and to avoid the effects of an illegal contract.

Conditions, warranties and innominate terms

Not all terms of a contract are of equal stance, and they have, thus, been classified into conditions, warranties and innominate terms. A condition is generally an important term of a contract and breach of a condition will entitle the injured party to be released from the contract and sue for damages. A warranty is less important than a condition, and breach of a warranty allows a right to sue for damages but not to be released from the contract. The third type of term is an innominate term, which can be classified as a more or less important term depending upon the effects of the breach. These are considered after the nature of the conditions has been examined; *Hong Kong Fir Shipping Co. Ltd v. Kawasaki Kisen Kaisha Ltd, CA, 1962.*

Implied terms

Where a term is not expressly stated, but is one which the parties, in the view of the courts, must have intended to include to give the contract business efficacy, the term may be implied into the contract. To be implied by the court the term needs to be 'something so obvious it goes without saying'. A court can also imply terms as a matter of law, although many terms are within statutes, for example the Supply of Goods and Services Act 1982.

Local custom or trade systems can imply terms to fill areas that a contract does not address; *Hutton v. Warren (1836).* A custom term cannot be applied if a contrary express term exists in a contract.

The third type of implied term is through statute. Statutorily implied terms are not based on the intention of the parties but on the rules of law and public policy. They should not give effect to the intention of the parties, but provide some protection for the expectation of the purchaser.

Exclusion clauses

An exclusion clause is inserted into a contract to remove or financially limit a party's liability for breach of contract, misrepresentation or negligence. The court uses various means to control these clauses, and the Unfair Contracts Terms Act 1977 and the Unfair Terms in Consumer Contracts Regulations 1999 exist. Control by the courts is generally run using 'incorporation' and 'as a matter of construction'.

By signing a document containing exclusion clauses, the party is bound by those terms regardless of whether the document has been read or is understood; *L'Estrange v. Graucob, CA, 1934.* When the document is unsigned and just delivered, then reasonable and sufficient notice of the exclusion clauses must be made, i.e. 'for conditions see back' notices on tickets; *Thompson v. L.M.S. Railway Co. (1930).*

Where there has been insufficient notice, exclusion clauses can still be incorporated where there have been previous dealings between the parties on similar terms; *Olley v. Marlborough Court, CA, 1949.*

Several pieces of legislation exist to control exclusion clauses in contracts, namely the Unfair Contract Terms Act 1977, the Fair Trading Act 1973 and the Unfair Terms in Consumer Contracts Regulation 1999, which is related to EC Directive 93/13.

Privity of contract

The doctrine of privity is closely associated with the doctrine of consideration. Upex (1999) states a general phrase linked to the doctrine of privity:

> A contract cannot confer benefits on strangers, nor can it impose burdens on strangers.

In essence, the rule of privity means a person who is not privy to a contract cannot sue or be sued under that contract. The most referred case is *Dunlop Pneumatic Tyre Co. Ltd v. Selfridge Co. Ltd, HL, 1915*, where the plaintiffs sold tyres to a wholesale distribution company, who then sold to the defendant. The defendant then sold these tyres at a reduced price lower than specified by the manufacturer/plaintiff and the plaintiff tried to sue for breaching a clause in the contract between the plaintiff and the wholesale distributors.

Following the idea that a contract is a bargain, even if a person is mentioned in an agreement and stated to be the promise, if such persons give no consideration they will not be able to sue. Though they may be party to the agreement, they are not party to the bargain and, thus, are not party to the contract. There are a number of statutory exceptions, for example the Law of Property Act 1925, s.136; the Bills of Exchange Act 1882, s.29; and the Married Women's Property Act 1882, s.11.

Two main exceptions to the rules exist. Firstly, when a person knows of a contract but is not party to that contract and induces a party to that contract to break it, the innocent party to the contract can sue the third person even though that person is not party to the contract. The second is where a restrictive covenant is made affecting land. Here a later owner of the land can enforce it even though they were not a party to the original covenant.

Agency is also a main common law exception to privity and occurs when one person (the principal) appoints an agent to enter into a contract on their behalf with a third party. The principal can sue the third party; *N.Z. Shipping Co. Ltd v. A.M. Satterthwaite, PC, 1975*. Along with agency, the doctrine of privity does not forbid assignment of contractual rights. One of the final relevant exceptional areas is collateral contracts, which can be used to evade the privity rule.

Another recognised method of evading the doctrine of privity is the trust device. This is where A contracts with B to pass on the benefit to C and B is holding the contractual rights for C in trust. C can then ask B to sue A as trustee, or if B relents, sue A themselves as co-defendant; *Les Afréteurs Réunis SA v. Walford, HL, 1919*.

Discharge of contracts

Contracts can be discharged through frustration, breach of the contract, agreement or performance of the contract.

A contract is frustrated where, after the contract was formed, events occur which make performance of the contract impossible, illegal or something radically different from that which was in the contemplation of the parties at the time they entered into the contract. Contracts discharged on the grounds of frustration are brought to an end automatically by the operation of a rule of law, irrespective of the wishes of the parties; *Hirji Mulji v. Cheong Yue (1926)*.

One party cannot, by a wrongful act, bring a contract to an end without the consent of the other. If A commits a breach of a condition, or a fundamental breach of an innominate term of a contract with B, it is for B to decide whether to terminate the contract or not. Breach of the main types of term has various methods of resolution. Some allow only damage claims, for example a warranty, whereas the more serious allow a discharge of contract along with damages, for example a condition, and are often called 'repudiatory' breaches.

A contract can be discharged or varied by agreement. This is where both parties give up their rights under the previous contract and either go their separate ways or provide new consideration for a new contract. Should this consideration result in a fundamental change to the original contract, then rescission should occur. If the implementation of change is small a variation can be utilised or one party can waiver its rights under the doctrine of waiver. If both parties perform their obligations under the contract, precisely and exactly, the contract is discharged; *Re Moore Co. & Landover Co. (1921)*. However, this has been limited recently by section 4(1) of the Sale & Supply of Goods Act 1994.

Contracts are enforceable if a party's contractual obligations have been met substantially, but not precisely. Where performance by one party is partial, the other party can accept this partial performance; *Christy v. Row (1808)*. When partial performance is accepted, the partial performer can claim *quantum meruit* costs.

Where one party is unable to complete performance without working together with the other party, the party may make an offer of performance which is then rejected by the other party. The party offering or tendering the performance will then be discharged from further liability. A tender of performance is equal to performance; *Startup v. Macdonald (1843)*.

If a definite time for performance is included within a contract, a question of essence is raised. At common law, unless the contract provides otherwise, performance must be completed by the date specified, otherwise it is a repudiatory breach. In equity, the courts are slightly more lenient and specific performance may be decreed despite the failure of the plaintiff to keep to the time fixed for completion.

Remedies

The remedies for breach of contract exist in common law, which provides the remedy of damages, and in equity, which provides the remedies of rescission, rectification, specific performance and injunction.

Damages

Damages are awarded to position the injured party in the same financial position as it would have been had the contract been performed correctly, to provide a compensation for the loss of the bargain within the contract. There are also *reliance damages* that pay for the costs incurred as a result of the defendant's breach and are there to return the plaintiff to that party's original position.

When a claimant has been negligent, with it contributing to the damage this party has suffered, the damages payable to the claimant can be reduced under the Law Reform (Contributory Negligence) Act 1945, except under a breach of strict contractual duty. There is also duty on the innocent party to mitigate loss, by taking reasonable steps to minimise the loss and not to make any attempts to increase the loss.

A plaintiff may not be able to recover the damages for all the losses suffered because, under the *Hadley v. Baxendale (1854)* ruling, some losses may be adjudged to be too remote from the compensation breach in question. This has resulted in a precedence whereby a plaintiff can only recover damages when the loss arises naturally out of the breach or could reasonably be supposed to have been within the contemplation of the parties at the time the contract was formed.

A plaintiff seeking to recover damages is sometimes said to be under a duty to mitigate, although the defendant has the burden of proving failure to mitigate; *British Westinghouse Electric Co. v. Underground Electric Railway Co. of London, 1IL, 1912*. The plaintiff is only required to act reasonably in order to mitigate and has no need to embark on hazardous or uncertain courses of action; *Pilkington v. Wood (1953)*.

A contract may contain a clause providing for the payment of a fixed sum on breach, and so long as the sum is a genuine pre-estimate of the loss, the plaintiff is allowed to recover this sum on breach without proof of actual loss. Such damages are often called 'liquidated damages' or 'agreed damages.' Standard forms of contract used in the construction industry generally acknowledge and include within the contract a clause for liquidated damages. If the sum stated in the contract is not a genuine pre-estimate then the payment is a penalty. Should the loss be greater than the stated penalty, then the full amount can be claimed, unlike liquidated damages.

Quantum meruit

A party can claim on a *quantum meruit* ('as much as he has deserved') where the contract makes no express provision for remuneration; for example, in the sale of goods, if the contract does not fix the price the buyer must pay a reasonable price. Another form where this payment (note, not compensation) can be claimed is when one party is prevented by the other from completing performance.

Rescission and rectification

Rescission is an order by the courts for the parties to a contract to be returned to the positions they occupied before the contract was made. Careful consideration of any innocent third party acquiring rights has to be given, and if it is not possible to rescind the contract, the matter will be dealt with using the award of damages. The equitable remedy of rectification is where a written contract does not fully express the true intention of the parties of the contract, and the courts have the power to rectify or alter the original document to reflect the true meaning of the agreement. Rectification is not a panacea for poorly written contracts and is only rarely used in practice.

Specific performance and injunction

These are equitable remedies awarded at the discretion of the courts. Specific performance is an order of the court compelling the defendant to perform his/her part of the contract, although this can be refused by the defendant on the ground of hardship; *Patel v. Ali (1984)*. An injunction is a decree by the court ordering a person to do or not do a certain act and in contract law is used to restrain a party from committing a breach of contract. It can be used to prevent a party breaking a contract and cannot force a person to do something that would be ordered by specific performance; *Lumley v. Wagner (1852)*.

Summary

This chapter has briefly explained the allocation of risk and its perception by the various parties involved with construction contracts. It then went on to describe the key elements in contractor selection, the role of pre-qualification and the criteria for contract award.

Bibliography

General

Smith, N. J. *Managing Risk in Construction Projects.* Blackwell Science, Oxford, 1999.

Turner, J. R. *The Commercial Project Manager.* McGraw-Hill, London, 1995.

Uff, J. and Capper, P. *Construction Contract Policy.* Centre of Construction Law and Management, King's College, London, 1989.

Law

Duxbury, R. *Contracts in a Nutshell,* 5th edition. Sweet & Maxwell London, 2001.

Great Britain. *Contracts (Rights of 3rd Parties) Act 1999.* Stationery Office Books, London, 1999.

Great Britain. *Law Reform (Contributory Negligence) Act 1945.* Stationery Office Books, London, 1945.

Great Britain. *The Sale and Supply of Goods Act 1994.* Stationery Office Books, London, 1994.

Great Britain. *Trade Unions and Labour Relations (Consolidation) Act 1992.* HMSO, London, 1992.

Institution of Civil Engineers. *ICE Conditions of Contract: Conditions of Contract and Forms of Tender, Agreement and Bond for Use in Connection with Works of Civil Engineering Construction.* Thomas Telford, London, 1999.

McKendrick, E. *Contract Law,* 4th edition. Macmillan, London, 2000.

Manson, K. *Law for Civil Engineers: an Introduction.* Longman, Harlow, 1993.

Murdoch, J. and Hughes, W. *Construction Contracts – Law and Management,* 3rd edition. Spon, London, 2000.

Smith, J. C. *The Law of Contract,* 4th edition. Sweet & Maxwell, London, 2002.

Uff, J. *Construction Law,* 7th edition. Sweet & Maxwell, London, 2000.

Upex, R. *Davies on Contract,* 8th edition. Sweet & Maxwell, London, 1999.

Wheeler, S. and Shaw, J. *Contract Law, Cases, Materials, and Commentary.* Oxford University Press, Oxford, 1996.

CHAPTER THREE

Drafting and using construction contracts – a legal perspective

E. Davies and C. Hallam

Introduction

There is a commonly held view that a construction contract is essentially a legal document which, once agreed and signed by the parties, should be 'put in a drawer' and not looked at unless there is a dispute. Those who hold this view may regard project management as a process that is largely disconnected from the contract, and may consider reference to the contract by the parties as adversarial and therefore undesirable. However, simply ignoring the contract may prove to be perilous. For example, the contract may contain certain 'conditions precedent' that have to be met before a party is entitled to bring a claim – if the parties have not read the contract, they will not be aware of those conditions. A project manager who has ignored the contract is likely to be more than a little distressed when told that a legitimate claim has been lost because the legitimacy of the claim depends on a notice being served by a particular date – and no notice has been served.

In this chapter, it is considered that the construction contract should to be used as part of the project management procedure, and that the use of the contract should be no more controversial than the use of well-established project management procedures. It is also suggested that, in the majority of cases, proper use of the management powers contained in the contract will prove to be more useful than reliance upon the legal remedies arising therefrom, such as monetary damages, termination and the like. In doing this, some ideas as to how to set up and draft contracts are explained.

This chapter is written mainly from the point of view of the organisation procuring the work (referred to as 'the client') because it is usual for the client (or its advisers) to prepare the contract. However, many of the points made apply equally to subcontracting. There is also reference to a person employed by the client to act as its representative in relation to the contract. This person is called 'the engineer', although it could equally be an architect or other professional project/contract manager.

The first point is that the contract itself should not be the starting point in the process. The contract should be a servant of the parties rather than their master – it should record, clearly and concisely, and in a manner that can be easily understood, the commercial agreement that has been reached by the parties. However three other documents should be created first. These are a *wish list*, a *constraints list* and a *risk matrix*. The purpose of these documents is to be the source material and a checklist for the contract.

Wish list

The wish list should set out what the 'operational team' – i.e. the people who are ultimately going to use the facility wished – to achieve. It is quite common for there to be a project team responsible for procurement (made up of employees of or advisers to the client) which is totally separate from the operational team. They may well come from different professions, backgrounds or experience. The first task for the project team is to understand the needs and wishes of the operational team and this is often more difficult than people imagine. There is an old saying, 'be careful what you ask for, you might get it!', and this should be borne in mind.

The wishes should be divided into three grades of importance, being:

- A – 'essential' items
- B – 'useful' items
- C – items that would be 'nice to have'.

The operational team must be prepared to explain why each item is in the list and why it is within list A, B or C. In order to focus the minds of the operational team, they should assume that if any of the items on the A list cannot be achieved then the project will be cancelled.

It goes without saying that it is desirable for projects to be completed within budget. One of the main reasons that budgets are exceeded is that the essential items are changed after the contract is let. Sometimes this is unavoidable: it may be that a particular component becomes unavailable and a more expensive alternative has to be used; it may be that the business needs of the client change; sometimes it is simply impossible to foresee all eventualities and changes are therefore unavoidable. However, it is worth spending a considerable time on the wish list to make it as firm as possible, as this document will ultimately turn into a specification to be incorporated into the contract.

Another issue arises at this point, which is how prescriptive the wish list should be. It is useful to think of 'design' as a series of decisions. In essence, the wish list is a series of decisions. At the top level, a decision needs to be made to embark upon a project. A company might decide that it needs a process plant; a public authority might decide that a road should be built. The next level down might be precise location and size. At the opposite end of the scale are more detailed (but potentially important) decisions about minor elements of the work. It has even been suggested that the question of how hard to hit a nail into a piece of wood is a decision in this sense.

The question is how many of those decisions should be taken by the client and how many by the contractor. This will depend on a number of factors, but possibly the most important is the nature of the project. In some cases the client will be very interested in a particular decision, in other cases not.

Take the case of a process plant. There might, for example, at one end of the site that is rarely visited, be a particular piece of equipment that needs to be inside a small building. The purpose of the building is to provide an element of weatherproofing for the equipment. Assume that the internal environment is not critical. It would not be important if there was the occasional leak or fluctuation in temperature; all that matters is that the equipment is, over a period of years, inside rather than outside. In that situation, a client might well be prepared for the contractor to design the building.

Another case is the front elevation of a museum building of national importance in a prominent location. Here, although there might not be major structural or engineering issues, the exact specification of the frontage will be of critical importance to the client. Although it might be acceptable for the engineer (or more likely an architect) or the contractor to suggest designs, the client will want to retain total control over the decision that is finally made.

Why is this distinction important? Any decision taken by the client (or on its behalf) will generally be the client's responsibility. If the decision is wrong in some way and/or needs to be changed after the contract has been let then there will generally be a cost implication for the client. However, a decision made by the contractor will generally be the contractor's responsibility, will be at the contractor's risk, and any cost implication will lie with the contractor.

A compromise might be to introduce an 'approvals' regime where a decision is provisionally made by the contractor but submitted to the client (or, more usually, the engineer) for approval. The approval is not an acceptance of responsibility for the decision, simply a permission to proceed. These regimes are usually constructed so that the approver is enforcing the contract rather than changing it, but there is a conceptual difficulty with an approvals regime. Take an extreme example. Suppose a contractor has to submit the proposed colour of the paint for a particular surface to the engineer for approval. Every colour suggested is rejected. The engineer says that only one colour will be approved, but it is very unusual and will cost a great deal more than a normal colour, and there is no engineering reason to choose this colour. It may well be possible under a particular legal system to write a contract that would allow such a thing to happen, but it seems logical that there must be some implicit element of reasonableness in an approvals regime, or at least a link to functionality. In the absence of clear provisions to the contrary, the approver cannot use an approvals regime as a tool to impose the approver's design decisions on the contractor. The rejection of something offered for approval must be based on objective and sound criteria in an effort to enforce the contract by, for example, rejecting something that will not perform its required function. The approver may ask for alternatives so that a choice can be made but there is a line which should not be crossed.

This is an issue commonly seen in disputes. Best practice is for the contract to contain a clear and concise approvals regime, which sets out the basis on which approvals will be given or withheld. Contractors should be wary of approvals regimes that are not clear in this way. In preparing the wish list, decisions which are to be left to the contractor but which are to be the subject of approvals should be thought through carefully. It may be that contractors will come up with good ideas, and leaving some of the thinking to them may help the tender process in that contractors will have the opportunity either to innovate or to demonstrate their ability to do so. If that is the policy then approvals should be a way of protecting functionality and perhaps a limited element of aesthetics. However, if the real aim is to take design decisions, then perhaps delegation of those decisions subject to approval will often prove problematic.

Constraints

The constraints list contains all the limiting factors which may affect the implementation of the project. These are factors which are external to the project and are actual constraints, rather than possible constraints, to distinguish them from risks, which should be contained in the risk matrix document (see below). Some typical constraints are set out below, but this should only be used for guidance, as the list of applicable constraints will be unique to each project.

Funding constraints

Possibly the most important constraint will be budgetary. There is usually a limit to the amount of money which is available for the project. This may be the client's own funds or money which is to be obtained from outside sources such as banks or other lending institutions. In commercial situations, the available funds may well be a variable depending on the details of the project. For example, in the case of a power station, the amount of electricity that can be produced (and therefore the potential revenue) will depend on the size of the plant and, in turn, the size of the plant will have an impact on the price. Thus there will be an influence from the wish list on the constraints list and vice versa.

Where finance is obtained from an external source, for example a lending bank, the bank will almost certainly impose its own constraints on the project. This is particularly so in the case of non-recourse financing, where the bank does not take security from the general assets of the client, but instead the bank's security is, for example, the income stream generated by the power station. In some cases the bank's requirements will be very detailed. They may require certain elements of the work to be procured in particular ways, for example the bank may require that, in respect of contractors, there is single-point responsibility for defects. They may also require that contractors and certain subcontractors enter into direct agreements with the bank warranting the quality of their work, and/or granting to the bank 'step-in' rights, to enable the bank to 'step in' and take over the project if certain circumstances occur, for example if the client becomes insolvent.

The availability of funding may well have an impact on the payment regime adopted in the contract. It will be essential to ensure that the liability to pay the contractor arises at such times and in such amounts as will be available from the funder. If the funder is only prepared to release money after certain stages of the work have been completed, then this would suggest that a milestone approach to payment in the construction contract will be necessary, rather than monthly payment pursuant to the contractor's applications.

In some cases there will be two funders, one for the construction phase of the project and another for the operational phase, and each funder may have different requirements. Typically, the operational finance might only become available after commissioning tests have been successfully completed. In that case it will be necessary to ensure that the requirements of the contractor under the construction contract match the requirements for the availability of the operational finance, as well as any requirements of the construction phase funder.

Given that funding may be conditional on complying with the bank's requirements, steps will need to be taken in any construction contract to ensure that these requirements are properly and timeously complied with. In many cases this is overlooked and clients have to rely on post-contractual negotiation with the contractor to provide such documents. It may not even be enough to place a simple obligation on the contractor to provide warranties in favour of the funder, as a question arises as to how that obligation would be enforced in practice. One should not assume that the courts will order a reluctant contractor to sign such a document – in many cases they will not and in some countries such an order is not within the courts' powers. It is not unusual to make the contractor's right to payment conditional on the provision of such documents. This can be justified if the provision of the documents is required to release funding and the position (and the documents involved) is clearly set out at tender stage. However, care should be taken to ensure that such a provision will be legally enforceable under the applicable law, as it may amount to an unenforceable penalty.

Regulatory constraints

In many cases, approvals from government or other regulatory bodies will be required before the project can proceed.

The most common is under planning legislation (known as zoning in some countries). This is a system where local government decides which areas under its control will be used for particular purposes, such as residential or industrial. Sometimes consent will be given for a project in outline but further, detailed consent will be required at a later stage. For example, outline permission might be granted for a structure which crosses a river by way of a bridge, but consent for the details of the portals through which the river will flow might be deferred. The person seeking the permission would not want to incur the expense of having detailed design work carried out until permission had been obtained for the scheme as a whole.

Usually the outline permission is obtained by the client, but it is becoming increasingly common for the contractor to be asked to obtain the detailed consent (and to take the risk for any failure to do so) where the subject matter of that consent is being designed by the contractor. A decision will need to be made as to who will obtain the consent, and this will need to be written into the contract.

There are many other types of consents that need to be obtained and these will differ from project to project. In all cases the exercise of the regulatory powers can have a significant impact on the wish list. Early consultation with the regulatory authorities is recommended to find out what is likely to be approved. Ideally, the consents should be obtained before the contract is let, and then any conditions attached to the consents can be incorporated into the specification. If this is not possible, a decision needs to be made as to whether the contractor should be asked to obtain the consent, in which case there should be enough flexibility or discretion afforded to the contractor to enable the consent eventually obtained to be complied with without having to change the specification of work with which the contractor must comply. Otherwise, under most contracts the change will be at the cost of the client (both in terms of capital expenditure and additional costs caused by programming delays).

It may be possible to provide that the contractor will bear the cost of a change necessitated by regulatory consents; however, it is suggested that this is not ideal if the contractor is being asked to bear a risk which the client is going to manage. Disputes may arise as to the conduct of the relevant applications.

Other regulatory constraints include applicable laws where no specific consents are required. Again, the relevant laws will differ from project to project and from country to country. These include laws relating to health and safety and environmental matters. Typically these are laws which must always be complied with and action will only be taken by the relevant authority in the case of breach. It is best practice to identify the most relevant laws in the construction contract (on a non-exhaustive basis) and to expressly require compliance with them. The contract can also require the contractor to comply with 'all other legislation' by a single 'wrapping up' provision.

Special care is required where components are being manufactured in one country for use in another. The regulations of the first country may require that the components are manufactured in a way which breaches, or at least is inconsistent with, the regulations of the second, or vice versa. Many countries have special rules to cater for this situation but a contract which simply requires that 'all applicable laws' should be complied with may run into difficulty. As ever, the contract should consider this constraint in detail and set out exactly what the contractor is to do.

Third party rights
In many cases a project will impact upon land, structures or equipment belonging to others. In the simplest cases the project will require that the land is purchased from the third party. In more

complicated cases there may be detailed arrangements for access and permanent arrangements for the sharing of land or equipment. For example, a pipeline may need to pass through a field owned by a farmer. Permission will be needed to enter the field and dig a trench for the pipeline. After the works are complete there will need to be a legal right for the pipeline to remain on the farmer's land. In some countries the pipeline will become the property of the farmer unless legal arrangements are made to keep it within the ownership of the pipeline company. Access may also be needed for maintenance purposes in the future.

Other projects (particularly urban regeneration) involve extremely complex arrangements, where land is used for different purposes at various stages in a project by different developers and contractors. There may be demolition of a building, the site of which is then used as a contractor's compound and then subsequently for further building work. The sequence of possession dates can have a major impact on the construction programme and it is common for the project to be divided into phases during which different areas of land and/or the site will be available to the contractor. Many contracts provide for differing access to the site for contractors at various points in the project.

It goes without saying that such arrangements need to be entered into at an early stage. There have been cases where pipeline projects have been held up because arrangements have not been finalised with landowners through whose land the pipeline is to pass. Similarly, the denial of access to a contractor to a part of the site which the contractor is entitled to use will usually lead to a claim for wasted costs being made by the contractor, and the project will inevitably be delayed, which will have other cost consequences for the client. If the precise timings cannot be ascertained in advance, arrangements may have to be made which involve giving notices to a contractor with pre-agreed mobilisation times.

Commercial and operational constraints

These are constraints that relate to the business or activities of the client. Timing constraints are typical of this category; for example, it might be that a project for the resurfacing of an airport runway can only be carried out at night, because the airport needs to use the runway during the day. Another example might be the need to wait for the availability of alternative accommodation for people working in a building that is to be redeveloped. A school may need to be ready for an academic term. Again, these are matters which will need to be discussed with the client's operational team and they can have a major impact on programming.

Although some contracts have acceleration measures in the case of programme slippage, the powers of the client (or the engineer) may need to be greatly enhanced if timing issues are critical. This is dealt with in more detail in the risk matrix section, below.

Other constraints may relate to the client. Many organisations have ethical policies which need to be complied with relating to materials or sources of supply. In the case of materials, the prohibited items might be those which could damage the environment or those from suppliers whose activities are not approved. There is a fair diversity of opinion on what is and what is not 'ethical' and this may differ markedly from country to country. Such policies, particularly if they go beyond the general law and current engineering practice, should be identified to the contractor at tender stage so that pricing can be carried out with compliance in mind. Some policies also relate to timing – it may not be acceptable to work on certain religious holidays – and these will need to be authoritatively identified and written into the contract. Even minor exceptions to the policy may require approval from the highest levels of management of the client, which can take a considerable time to obtain.

Supply and labour constraints

The project may require materials, components or specialist skills which are in short supply or need long lead-in times. This may require the letting of advance contracts to secure such supply. There may also be political issues such as delays and problems in obtaining visas and work permits in international contracts.

Risk matrix

The third document that needs to be prepared is the risk matrix. It is suggested that this is drawn up at a 'brainstorming' meeting attended by the whole project team. The usual rules apply (no idea is ridiculous, etc.), and this meeting is often quite good for team 'bonding' purposes if the people do not know each other well. Each risk is assessed for:

- high/low/medium probability
- high/low/medium financial impact
- high/low/medium time impact.

It should be considered whether or not there is an ability to 'lay off' the risk to insurance or the contractor, whilst bearing in mind the general principle that a risk is usually minimised if taken by the person best able to manage its causative factors. There is also the issue of loss of control of risks managed by contractors, which can sometimes be controversial.

Generally, if a risk is taken by the client, the cost consequences if the risk materialises will be an actual cost. If a risk is laid off to insurance or to a contractor, the client will still pay for that risk but the cost will be a statistical cost. It is for this reason that some organisations 'self-insure' and take the risk themselves.

For example, suppose there is a risk that a particular component will be damaged during installation, regardless of how carefully it is installed. If the client bears the risk and the component is damaged, the cost will be the repair or replacement price of the component, and the cost of removal and reinstallation with attendant delay. However, if the component is not damaged, the cost is nil. If the risk is passed to insurers or a contractor, they will make an assessment of the risk and add this to their price or premium as appropriate. The client will pay this cost whether or not the component is damaged.

Obviously there is a balancing exercise to be carried out, but it should be borne in mind that a risk laid off is not necessarily a cost saving in the long term, particularly if the client is better able to manage the risk in question.

The next part of this chapter explains, with examples, how contracts can be used as a risk management tool.

Defective work

One of the most important risks for the client to consider is the risk that the contractor's work will be in some way defective. The first step to be taken in contractual terms is to ensure that the work that the contractor has to do and the standards to be achieved are clearly set out in the contract. One often sees words such as 'good', 'proper', 'workmanlike', 'fit', 'suitable' or 'appropriate' used. The difficulty with those words is that they are subject to interpretation. If the contractor regards something as 'good' but the client disagrees it can be quite difficult to decide who is right, particularly in a borderline case, and perhaps more so where the contractor can be considered an expert but the client is merely a layperson. Although ultimately there should be a system of dispute resolution in the contract, referral to that system is perhaps best avoided, especially in the first instance. The resolution of the dispute may well turn on which party's expert witness is preferred by the dispute resolver.

Generally, the clearer the specification of work, the less room there is for dispute. Bear in mind that the word 'specification' starts with the word 'specific': if there is a recognised standard against which the work can be objectively measured, this should be expressly incorporated rather than risking a dispute as to whether it is applicable. Testing regimes should be as comprehensive as possible.

In some cases even more sophisticated measures can be adopted, such as having independent testers available to 'referee' any dispute as to compliance within a space of hours rather than days or weeks. This is particularly useful if work has to be covered up and if it would be expensive or impractical to uncover it later. Such arrangements typically would not absolve the contractor from liability for 'latent' defects. The tester needs to enter into a contract with both parties warranting the quality of testing. Many testers would, however, limit their liability to the use of reasonable skill and care rather than guaranteeing that the testing will be correct.

Most contracts will contain a power for the engineer to order the removal of defective work, but a difficulty arises where there are numerous problems and rectification work is continually unsatisfactory. There is a risk that a contractor (or a subcontractor) will not be able to produce the required standard or that to do so will take so long as to cause unacceptable delays, including delays to follow-on trades. Clients should ensure that the contract contains procedures which will enable a contractor to be removed from the project in this situation and/or for the relevant work to be carried out by others.

Procedures of this type should be used carefully and a client should be certain that there is enough evidence to justify the action taken in the event that there is a dispute. This is particularly important where the dispute turns on the interpretation of the specification. In some contracts there is a right for the client to terminate 'for convenience', in other words, without relying on some default on the part of the contractor. The consequences of this method of removing a contractor (for example, the payment terms) are usually more generous than those relating to removal in a situation where the contractor is in default. However, these provisions may prove to be cheaper for the client than the payments due where a termination for default has been successfully challenged by the contractor.

A subject of recent interest in English law has been the extent to which a client can enforce the specification. Suppose a swimming pool is constructed slightly shallower than required by the specification. This may only amount to a few centimetres but the cost of rectifying it may be as much as the swimming pool cost to construct in the first place. This scenario was heard in the English courts in the case of *Ruxley v. Forsyth*, where the client contended that the contractor had failed to construct what he had agreed. The court decided that it would be unreasonable to require the contractor to carry out the rectification work (i.e. to reconstruct the swimming pool to the required depth) and that a payment of damages for diminution in value of the swimming pool was the appropriate remedy. The measure of damages was relatively minor – after all, does it really matter that a swimming pool is few centimetres shallower than required by the contract? This has potential consequences for the enforcement of specifications in many instances, especially where it is argued by the contractor that the works have been 'over-engineered'. To design in such a way may be the deliberate policy of the client. The lesson to be learned is to focus on and provide procedures for inspection, testing, quality control and rectification of defects at the time of construction (when, in theory at least, repairs ought to be less expensive to carry out) rather than after the event.

If variations are unavoidable, a system of pre-pricing can be useful, particularly if it includes an agreed entitlement to extensions of time and payment for delay and disruption. However, it should be borne in mind that the pre-pricing process does not match a competitive tender and will not always ensure value for money. It is suggested that the client should have the right to order the variation on an 'actual cost' basis by reference to a valuation system in the event that the 'pre-price' cannot be agreed. Alternatively, if certainty is required, the pre-pricing could be delegated to an independent third party (to be completed prior to the instruction being given, if appropriate) in the case of failure to agree.

Delay
Delay is another risk often uppermost in the client's consideration. Traditionally, the threat of liquidated damages has been considered sufficient 'incentive' for a contractor to finish on time. However, the causes of delay are commonly a matter of dispute, which is often only resolved after completion. During the project, the contractor has a dilemma as to whether to institute accelerative measures in the absence of an instruction to do so from the client. If, at the end of the project, it is finally decided that the delay was principally the fault of the client or the engineer, it is by no means clear in many contracts (and in some countries even less clear under the general law) whether the costs of such measures are a legitimate claim against the client. On the other hand, if it is decided that the delay was the responsibility of the contractor (and there is therefore no entitlement to an extension of time), it may have been cheaper for the contractor to accelerate the works than to pay liquidated damages. Of course, most delayed projects are delayed by a number of causes, with fault on both sides, and this makes the position even more complex.

If time is critical, then clients should consider acceleration procedures which go beyond the usual (and often rather vague and limited) powers contained in standard forms. In some projects it might be possible to agree in advance what acceleration measures might be available in the case of delay (regardless of who is responsible for the delay). This might include overtime working or double shifts, the bringing onto site of extra plant or resources or a change in working methods. The client could have a power to order the implementation of such measures (with special payment arrangements), with the cost being met by the party responsible for the delay once such responsibility has been ascertained. In this way, the contract focuses on solving the problem at the time rather than being used as a vehicle for obtaining compensation after the event.

Liquidated damages for delay can be problematic if it is difficult to make a genuine pre-estimate of the loss that would be suffered by the client in the case of late completion. Damages which exceed such an amount – typically those designed to 'incentivise' the contractor rather than to compensate the client – are likely to be unenforceable in English law. In some countries liquidated damages themselves are unenforceable without proof of equivalent loss.

Insurable risks

Insurance is seen by many as a complex area best left to those who specialise in it. This attitude can result in a disconnection between the insurance cover arranged for the project and the risks apportioned by the contract between the parties.

One area of particular interest is the question of subrogation. This is a process where an insurer can recover money from the person or organisation who caused the loss which led to the claim. Take the example of a fire caused by the actions of a subcontractor. The client instructs remedial work, which is funded by a claim on the client's insurance policy. The insurers may in certain circumstances be able to recover that money from the subcontractor. This means that the subcontractor also needs to take out an insurance policy. In order to avoid this situation, the parties may agree (and the contract may provide) that the insurance will cover all participants to the project such that no subrogation will take place. Naturally, this makes the insurance cover more expensive. There have been problems under some standard-form contracts which left open the question of whether this arrangement applied even where there was demonstrable negligence on the part of the subcontractor, and litigation ensued.

A further complication of this type of arrangement is the level of cover obtained. Sometimes a client is only obliged to insure for the direct cost of the remedial works. However, the consequences of a fire are not limited to those costs. There might be disruption to other parts of the project and delays. Who will bear those costs? Additionally, the client may want to instruct accelerative measures to mitigate such delays. Who pays for that? Insurance may be available for such costs and it should be matched with the procedures and powers set out in the contract.

Constraints

Many of the constraints mentioned above will develop into risks if they are not definitively resolved prior to letting the contract. Sometimes this is unavoidable if, for example, design work needs to be carried out for consideration by a regulatory authority before a consent can be given.

Early warning systems

The use of early warning systems is to be encouraged. This is where a contractor is obliged to bring to the client's attention any matter which could delay or disrupt the project or if there is any claim to be made. The idea is for the client or the engineer to have the opportunity to deal with the matter before it becomes a major problem. However, compliance with such clauses is often seen as too adversarial and there are frequent problems with them not being complied with. This issue is at the heart of the 'contract in the drawer' viewpoint mentioned above.

Choice of contract

As can be seen from the above, the parties should start by considering the needs of the project and only then go on to choose or draft a contract. The three processes set out above should produce clear objectives against which a choice of contract should be made. There are a few more points to bear in mind.

A standard-form contract with which participants in the project are familiar is preferred by most contracting parties. One of the reasons for this is that the procedures will be known and understood and may well be adopted to a greater or lesser extent into the parties' own internal procedures. Systems that are known and understood are more likely to be complied with and to be commercially acceptable to the parties.

It is always possible to amend a standard form; however, to do so can be difficult and is sometimes discouraged by the publisher. Some forms set out a particular risk allocation (e.g. the right to make a claim in respect of unforeseen ground conditions) in more than one clause. If only one of the clauses is amended, uncertainty can result. Some standard forms are very complex and full of inter-linked cross-references which are carried through into standard subcontracts. If a clause in the main contract is amended, references to that clause in the prescribed subcontract may become meaning-less and this will lead to uncertainty.

Many standard forms contain provisions (or lack provisions) because of the particular features of the projects for which they are intended. Using a form intended for one type of project in an alto-gether different type of project can be problematic because of either inappropriate provisions or the lack of important provisions, or both.

Bespoke forms take a great deal of time and effort to produce. They will be unfamiliar to the parties, and their provisions will not be tried and tested. However, if circumstances dictate, sometimes they may be the best option.

Summary

This chapter addresses some of the issues that need to be considered when using contracts in construction projects. Although contracts can provide a good 'fall-back remedy', their first use should be to manage situations that arise to the benefit of the project.

The views set out in this chapter are in summary form only and are not intended to be legal advice. Please take detailed advice before taking any action (or inaction!) in relation to actual projects.

CHAPTER FOUR

Procuring the services of a project manager

P. Spring and S. Wearne

Introduction

This chapter states how a project manager should be selected and what are suitable terms of appointment. The principles set out apply whether the project is large or small, and whether the project manager is an employee of the client or is a consultant hired for a particular project.

Project management and the project manager

The primary task of project management is to deliver the maximum value in return for the resources employed. That usually means delivering a project to an agreed quality, safety, time and cost.

To achieve all this depends upon appointing a person as project manager who is dedicated to delivering the project. This role can be a separate job. It can be part of the tasks of someone working on the project, depending upon the size of the project, its risks and the competence of everyone, but in all cases the role is crucial through the planning and control of every project.

It is the role dedicated to the objectives of the project which is essential. Titles for the role vary. 'Project director' is appropriate for a project vital to an organisation's future. Weak titles such as 'project coordinator' or 'project engineer' indicate weak intentions. 'Project manager' is preferable for most projects, as it acknowledges that the tasks are managerial and shows the organisation recognises the value of appointing one person to represent its interests in achieving a successful project.

Research confirms the authors' personal experience that success in the role depends more upon human skills than on expertise in the content of the project. The project manager needs skills in how to manage. No one person can be expert in all the range of work packages for the project. What may be valuable for the project manager is experience in the critical or most risky content of the project so as to be able to foresee problems and know the questions to ask. 'What questions to ask' is the fruit of all experience. It also establishes credibility in leading the team; engineers, for instance, prefer leaders who are engineers.

Requirements for project success

Before a client ever thinks about appointing any of its team, it needs to decide what it is looking to achieve from the project or assignment. Fully understanding and establishing the key success factors of any project is the most important stage, and that must be the first step.

The next step is to understand the skills required to deliver those success factors and, in turn, a successful project. There is no point in employing resources without fully understanding why they are required in the first place. This can be achieved by carrying out an initial SWOT analysis; this in turn will identify the real risks and the weaknesses in the team. There is nothing more certain than that, in the real world of project delivery, the weaknesses will reveal themselves or come back to plague the project.

Such assessments need to address the soft and hard skills. The softer skills of personality, age, mental wavelength and social compatibility are becoming more and more important. Can this person get the best out of this situation regardless of their harder technical skills?

All clients must apply the same means test to themselves because it is imperative for them to fully understand how knowledgeable they are and, particularly, to understand their own shortcomings. Are you an educated client with plenty of experience in your field, is it a new experience or is it that you are just short of time and resource?

It is very much a gap analysis exercise, because the project manager should be the client's greatest ally and friend whilst at the same time covering for the weaknesses of the team and the client in particular. The adage of 'you are only as strong as your weakest link' applies equally to managing projects.

Therefore the qualities required are a vital consideration. Experience shows that the best results come when the project manager is independent and does not have a dual role. As an example, architects have found great difficulty in handling multiple roles within construction projects.

The qualities required are generally ones of leadership, reliability, authority, integrity, practicality, open-mindedness; the person must be flexible but decisive in a democratic and supportive way rather than by pure dominance. These skills are in addition to the obvious relevant technical skills.

The first appointment in any project should be that of the project manager, even if only internally, because that person, in conjunction with the client, must provide the essential direction, discipline and drive. It is a common mistake to involve other disciplines first; this causes difficulties of undermining the authority of the project manager because a direct link has already been forged between the client and the early-appointed consultants. A project manager with experience, enthusiasm and energy for the project is a good starting point.

Formal appointment

The project manager should be appointed formally so as to be established as representing the project sponsor. This is important for the project sponsor even if the person designated is already an employee, as the role can require leadership on internal questions as much as in external relationships.

Terms of appointment

Standard Terms for the Appointment of a Project Manager are published by the Association for Project Management (1998). The terms are in two sections. The first section sets out the basic terms of employment of a project manager. This is not required if the project manager is already an employee.

The second part of the terms is a job description. It is a schedule of duties. Alternative schedules of duties are provided. Schedule S should normally be used, especially for small projects. Schedule C is for large construction projects. Schedule I is for IT projects, and Schedule M for manufacturing projects. These alternative schedules vary only in detail.

The standard terms should be used without alterations or additions. Alterations and additions to standard terms are a habit in some UK organisations, despite a lack of evidence that these have improved the results. Alterations and additions lead to the risk of making the terms inconsistent or incomplete, and they divert time and commitment from the project.

Detailed terms published by some public organisations should be avoided. The task of project management is basically to do or get done what others are not doing that the project needs. All that can be forecast should be designed out by the initial attention to the objective, scope, risks and priorities of the project. Terms of reference which attempt to specify all the actions which might be required of a project manager are so lengthy that there is not time to read them and also do the job.

Terms of payment and risks

Provisions are made in the *Standard Terms for the Appointment of a Project Manager* for payment to the project manager on the basis of fixed prices, time charges and/or reimbursement of expenses. Fixed prices and time charges are normally considered to be alternative terms of payment, but they can be combined.

Agreement on fixed prices is not recommended before the project has been defined in sufficient detail to provide a quantitative basis for estimating the amount of work to be undertaken by the project manager. Convertible terms of payment may thus be appropriate, initially based on time charges and changing to fixed prices at a defined stage of the project.

The terms of payment chosen should be stated when inviting offers of services from project managers who are not already employees. The *Standard Terms* provide for agreed amounts to be inserted in a fee schedule.

The *Standard Terms* provide for agreed definitions of the limits to the project manager's authority, particularly spending authority, and requirements for professional indemnity insurance.

These provisions for payment and insurance are of course not required if the project manager is already an employee.

Pre-qualification

After deciding what is required of the project manager, it is then necessary to formalise that information and put it in an order that maximises the chances of selecting the most appropriate party for the job, whether relating to the individual or the organisation. The collecting and ordering of those all-important thoughts and defining processes are essential.

This is normally delivered through a briefing document that defines the facts about the project and, more importantly, defines what is expected from the project manager. This document should be quite detailed in terms of defining the areas of responsibilities and any particular requirements the client may have. As an example, it could indicate the requirements or expectations in terms of meetings, reports and essential checks and balances to meet the client organisation's needs.

At the same time, it is important to allow your candidates to express themselves and show how they can contribute to the success of the project. The object is to select a project manager with initiative. Give them the opportunity to demonstrate they have understood your requirements and the key issues and then go on to show how they will deliver. This is an important process because the client will gain from the expertise of all the bidders rather than just the selected bidder. Pre-qualification is a two-way street inasmuch as it is also an information and knowledge-gaining time for the selectors.

Traditional practice in most sectors of business and public services does not allow bidders to demonstrate their skills, and selection more often than not is just a function of the lowest price. In particular, this applies within the public sector, where there is not a meaningful accountable evaluation of value for money.

How to select the most suitable candidates is always challenging because, invariably, if there has been insufficient preparation then the chances of a successful selection are reduced.

The sourcing of suitable candidates is carried out in various ways. Invariably, the type of project will have been carried out before. Therefore it is not unreasonable to seek advice from colleagues in that line of business, perhaps via a trade or professional association. There are then project-management-specific organisations such as the Association for Project Management, The Institute of Management, The Institute of Directors or the local Chamber of Commerce. Such organisations bring the essential ingredients of independence, impartiality and clarity of the requirements.

Depending upon how competitive the industry sector is or how knowledge-sensitive the project is, then it is sometimes possible to ask a competitor if they can recommend organisations and learn from their experience.

Traditionally, the process starts with a long list of, say, six to ten names if it has not been possible to find a suitable candidate by chance. Now is the time for doing homework and checking out what is known about each of them. This can be achieved by an initial expression-of-interest-type document that is concise by nature. The objectives of such an exercise are to establish the experience, availability, suitability and details of the individuals and the organisations concerned.

From this exercise, a short list is prepared by analytical means. As an example, there is no point in appointing the most qualified person if that person is already committed to another project. The object of this short-listing exercise is to arrive at, say, no more than five similar candidates who are all capable of doing the job.

Depending upon the importance of the project, it may well be necessary to interview and take references as part of the short-listing process. However, this is tremendously time-consuming and generally deemed unnecessary prior to short-listing.

The main focus is on selecting the right individuals, but it is important not to lose sight of the type of organisation also being selected, particularly in the light of any collateral warranties or contractual guarantees that may be required to make the project work. Therefore the period in business, professional standing, size and status, financial stability, and level of resource can be key considerations in certain circumstances.

Selection and award

After the short list has been prepared, the respective project managers are invited to make further submissions in detail, with the focus being particularly on the needs of the project. Such topics for the candidates' consideration are the identification of the key issues and their proposals for their management and delivery.

This process needs to be undertaken in the knowledge and receipt of the latest information about the project, including an update of the brief, the scope of works, performance specifications, roles and responsibilities, appointment documentation requirements, risks, contractual requirements, programme, budget, and procedural issues. Projects are invariably continually evolving, and therefore this is a reiterative process with the aim of getting closer to the objectives at each stage.

Typically a project manager's presentation and interview lasts for an hour and a half, with the time being split between the presentation and time for questions and discussion.

Prior to commencing the interview, it is important to assemble your interview team. This needs to consist of the stakeholders, with the odd plant as an independent observer to throw in the 'curved ball question' to check out the candidates' flexibility and understanding rather than their delivering by rote. It is important not to be caught out by a bidder's 'A Team' who go from interview to interview but are never seen on the project.

Obtaining a buy-in from all the stakeholders is essential, so careful thought needs to be given to who they are and ensuring that they participate in the selection interviews.

Preparation is all. It is essential to take time to prepare. Make sure there is a structured score sheet containing the vital success factors and key skills requirements. These can be weighted against the key criteria, after taking time to discuss them and prioritise them in advance.

After the interviews the team should follow up immediately, preferably the same day, to score the candidates as a team whilst the issues and the performances are fresh to the mind. This gives an initial indication, but unless there is a truly outstanding candidate or the need to appoint is urgent then do not make a decision there and then. Take time to think, and most certainly make sure time is taken to talk to the key references at the earliest opportunity. It is surprising how one-sided a presentation can be without the recipients even realising it.

After taking the references on the further short-listed candidates, revisit the key-criteria schedule and re-mark it in the cold light of day. A golden rule is that project management is a people business; invariably, people are more important than organisations.

There is always a personality and character factor. In the real world, events do go awry. Hence every project has its difficult moments. This must be kept in mind in the selection process. It is essential to have the best person alongside when the going is tough and difficult decisions have got to be made.

Make your selection as quickly as possible without undue haste. Tell the successful candidate first. Invite the person in immediately to review and complete the appointment documentation, because it is again far easier to formalise the arrangement whilst it is fresh, rather than use a letter of intent or instruction initially and then try to follow up with the documentation months later.

If there is a concern about inadequate information to complete the appointment then it is time to be brave and appoint for an interim period, ensuring that the long-term interests are safeguarded. Attention to detail is essential at this stage.

Upon appointment, it is essential to advise the interested parties immediately and formally by stating the project manager's role and responsibilities, output requirements, and authority levels. It is amazing the number of clients who carry out a selection process and then do not have the courage to back the appointee. It is essential to empower the appointee and reinforce him/her to ensure the maximum chance of success.

The appointment is just the beginning and therefore, just as with all parties, it is essential to review performance standards on a regular basis. In the early phases this should take place every three months, because it is a major failing of projects that performance is only reviewed after the project has been completed. The review should be against the measured criteria set down in the appointment, adjusted to meet the changing demands along the way. The difficulty is in determining performance criteria that align the project manager with the client, rather than driving a wedge between them. The focus should be on incentive and reward rather than penalty.

Forward thinking

There is of course a downside to selecting a project manager, because the sceptics see such an appointment as an opportunity to take a back seat, with the common perception that it is the project manager's job to tell them what to do. Empowerment of all the team is essential, and keeping a watch in those early days to ensure the whole team is empowered rather than just the project manager is an essential act.

The whole object of having the team is to get the most out of all of the members. The 'dream team' is obtained when they are all star performers rather than one superstar with the rest being just mediocre.

Equally, the project manager is also looking for reassurance. Just as with any other member of the team, it helps to catch project managers doing things well and praising them for doing so. On the converse, if they are missing the point or the target then that must be brought to their attention promptly in a positive, analytical and factual manner.

The latest thinking is that perhaps the demands on the project manager are too much for one person, and it is becoming common practice to have a project management team consisting of the key stakeholders, who make joint decisions in the best interests of all parties and the project.

Completion of the service

The final task of the project manager is to review what went well on the project, and why, and what were the problems and their remedies. This closing review is a duty of the professional project manager.

Conclusions

The success of a project depends upon defining its objectives and priorities, assessing the risks and opportunities, and empowering a capable project manager to achieve these objectives. Selecting the project manager should likewise be based on defining the needs of the job and using clear and workable terms of appointment.

Bibliography

Association for Project Management. *Standard Terms for the Appointment of a Project Manager.* APM, High Wycombe, 1998.

Association for Project Management. *Project Management Body of Knowledge.* APM, High Wycombe, 2000.

Blackburn, S. The project manager and the project network. *International Journal of Project Management,* **20**(3) (2002), 199–204.

El-Sabaa, S. The skills and career path of an effective project manager. *International Journal of Project Management,* **19**(1) (2001), 1–7.

Sotiriou, D. and Wittmer, D. Influence methods of project managers: perceptions of team members and project managers. *Project Management Journal,* **32**(3) (2001), 12–20.

CHAPTER FIVE

Contract strategy

D. Bower

Introduction

Construction projects go through several stages from conception to completion, the actual construction phase is only one part of the overall process. Contract strategy is one of a series of decisions that are made during the early stages of a project. It is one of the most important decisions facing the client. The chosen contract strategy and the allocation of risk, the project management requirements, the design strategy, and the employment of consultants and contractors are inextricably linked and therefore the contract strategy has a major impact on the timescale and ultimate cost of the project.

Because of the diversity of both construction and the client's requirements, no single uniform approach to contractual arrangements can be specified or advocated. A number of alternative strategies are available to the client and each contract should be formulated with the specific job in mind. This chapter describes the role of the client in developing the contract strategy and describes the main procurement routes available.

The contracting environment

The construction project is often the result of additional demand, a need for replacement of certain facilities or the identification of an opportunity by a client who subsequently becomes the owner of the finished product; some projects extend to repair and maintenance. The client prepares the project brief, which sets out the project objectives and other basic requirements. Having set out a general idea of what is required, the client then consults a design team (sometimes, the design team is directly employed by the client), which develops an outline design of the required facility. The team may also be responsible for carrying out estimation of project costs. These would, however, be subject to the client's approval. More often than not, architects have led the design team, which may include quantity surveyors, structural engineers and services engineers. Civil engineers would often lead the team for civil engineering construction projects.

Once the outline design and the budget are approved by the client, the design team is ready to carry out a detail design and prepare detailed drawings and specifications. The design and specifications often go through several changes and modifications. The team also prepares tender documents that set out what the contractor is required to do, with penalties imposed for failing to deliver.

A number of contractors and, in turn, their suppliers are invited to bid for the construction works against the drawings and specifications, traditionally on the basis of a bill of quantities. This is often prepared by a quantity surveyor and it contains fairly complete specifications of the required work, as well as a schedule of rates for the works. Thus construction companies often prepare tender bids from inadequately detailed information. In most construction projects, this is the first time – the tendering phase – that the contractor is brought into the project.

In many traditional cases, especially in public works, the lowest complying bidder ultimately becomes the successful builder selected for the works. There is a tacit understanding that this situation prevails because of 'public' accountability. Knowing that selection is on a lowest-price basis and that changes during construction are inevitable, contractors and suppliers initially bid below cost to win the contract, but then they find it easy to raise the price because of the changes to specifications during the works. Once a builder's bid price is accepted, then that company is formally contracted to undertake the construction work as main contractor.

The successful construction company then plans out the construction works, which involves selecting a site team, establishing a detailed works programme and resource schedule, placing orders for materials and equipment, and arranging subcontracts with specialist companies. The planning is usually carried out by an office-based team of planners, buyers, surveyors and engineers. There is very little involvement of site workers in the planning process, and that is so even at a later stage when works have actually begun on site. However, on site, it is often the site workers who take full responsibility for the management of the works in accordance with the project plan, liasing with head office staff and members of the design team as deemed appropriate.

During construction, the main function of the main contractor is often to control the costs of its suppliers in order to maximise its own profit. Some of the parties within the construction process, particularly most suppliers and some subcontractors, are hardly 'remembered' on completion of the works. But generally, once the works are finished, the site team is often dissolved and a new team is established on a new project.

Designing and constructing or rehabilitating a facility is rarely straightforward. It is subject to a series of risks and uncertainties and involves a number of organisations specially assembled for the project. The way in which the client, the project manager and the various designers, contractors and suppliers work together as a team is determined by the form of contract entered into between the project participants and the client. The contract arrangements should be in accordance with the objectives of the project and should enable the risks to be controlled to achieve a successful outcome.

The role of the client

The ultimate responsibility for the management of a project lies squarely with the client; consequently, any project organisation structure should ensure that this ultimate responsibility can be effected. This is not to say that the client must be involved in the detailed project management but it does mean that the machinery must be in place for him to make critical decisions affecting his investment promptly whenever they become necessary.

The temporary nature of the project team and the need to define and achieve specific objectives against a demanding timescale, together with the high level of risk and expenditure encountered on many projects, will demand a positive style of project management.

The project manager must ensure that the client organisation supports the project team with direction, decision and drive and that it regularly reviews both objectives and performance.

The project manager

Projects that are technically complex or large in value will normally justify the appointment by the client of an individual or team whose sole task is that of project management. To be most effective, this appointment should be made early in the process, as soon as the proposed project receives serious consideration. The individual or group should thus become responsible for overall control of the project from its inception through to final commissioning. Whenever possible, the project manager should be an appropriately qualified professional or professionals, experienced in the type of project, and able to assume overall responsibility for a full range of necessary tasks. The approach will be most effective if there is continuity of service from inception to completion.

The role of the project manager is to control the evolution and execution of the project on behalf of the client and will therefore require a degree of executive authority if the coordination of activities is to be effective and progress maintained. Ideally, the project manager should be involved in the determination of the project objectives and subsequently in the evaluation of the contract strategy.

Project objectives

Construction projects are often complex with potential for cost and time overruns or the finished facility performing less well than planned. To minimise such risks, the client should select the contract strategy that matches the objectives of the project. These must be clearly established and prioritised before any design or other work begins.

The client must decide the relative importance of the three main objectives – time, cost and performance – of the completed project:

- *Time* – earlier completion can be achieved if construction is started before the design is finished. The greater the overlap between the two, the less time will be required to complete the project, but the amount of variation is likely to increase.
- *Cost* – with the exception of certain 'design and build' contract strategies, a final construction contract sum cannot be established with any degree of confidence until the design is complete. Any overlap between design and construction means that construction starts before the cost is fixed, increasing the uncertainty over cost.
- *Performance* – the quality and performance characteristics required from the completed facility determine both project time and cost. Some strategies reduce the clients' ability to control and make changes to the detailed specification after the contracts have been let. Performance includes the function of the facility, and its quality, appearance and durability, together with reliability and efficiency of operation.

The relative importance of each objective must be given careful consideration because decisions throughout the project will be based on balances between the other objectives. In a well-managed project the three objectives of time, performance and cost should be in constant tension. Usually an improvement in one can only be achieved at the cost of another. Too many projects, in both public and private sectors, overrun on time or cost, or underrun on performance because the manager does not keep all three objectives constantly in view. Clearly, if tight targets are set for all three objectives, the likelihood of meeting them all is small. Appropriate contingencies, in terms of tolerances in the specification, float in the programme and/or allowances in the budget, should therefore be allocated.

The impact of other secondary objectives should be minimised, as they frequently conflict with the primary objectives. Secondary objectives might include client involvement in management, generation of employment, early involvement of the contractor, use of capital and an early knowledge of actual cost. On overseas contracts they may also include the overlap of design with construction, the incorporation of appropriate technology and the provision, maintenance and disposal of plant. Once the project objectives have been set they must be communicated to all parties and, most importantly, adhered to.

The options for project organisation

The interrelationship of the basic requirements of contract strategy to balance flexibility, risk allocation and incentive with the type of contract is described below:

- *Incentive* – the aim is to provide an adequate incentive for efficient performance from the contractor. This must be reflected by an incentive for the client to provide appropriate information and support in a timely manner.
- *Flexibility* – the aim is to provide the client with sufficient flexibility to introduce change that can be anticipated but not defined at the tender stage. An important and related requirement is that the contract should provide for systematic and equitable evaluation of such changes. The introduction of changes must always be strictly controlled and limited to those that are essential for the safety and functioning of the facility.
- *Risk sharing* – the aim should be to allocate all risk between client and contractor in a realistic and equitable manner. This must take account of the management and control of the effects of risks that materialise. The contractor will include a risk contingency sum in its tender as protection against the risks it has been asked to carry. This contingency is hidden in the tender prices and is frequently inadequate if the lowest bid is accepted; this may consequently give problems if the risks materialise.

It is apparent that, generally, contractor's incentive and client's flexibility tend to be incompatible. For example, a lump sum contract imposes maximum incentive on the contractor but also implies a very high level of constraint on the client against introducing change. The converse is true at the other extreme of a cost-reimbursable plus percentage fee contract.

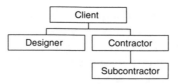

Figure 5.1. Traditional organisational structure

Organisational structure

A variety of organisational structures are available; in practice, some organisational structures are closely linked with a particular type of contract, for example the traditional approach with the admeasurement contract. As this is not always the case, it is preferable to consider the decision on organisational structure as separate from, but interrelated with, the decision on the type of contract.

Every time an interface is introduced into the project organisation the management effort required to deliver a successful project is increased, as is the risk of failure. The aim should be to minimise the number of interfaces between the different organisations. The organisational structure must define communication and contractual links.

Traditional

The traditional civil engineering contract is based on the separation of design and construction, as shown in Figure 5.1. The scope for involvement of the client in management and construction is limited, as is the contractor's input to the design of the permanent works.

The client, usually after carrying out a feasibility study, appoints a consulting engineer to undertake the detailed design, preparation of tender documents and site supervision duties. This appointment may be based on fee competition.

The client then appoints a contractor, usually through a competitive tendering process, to organise a detailed works programme and carry out the works. Ideally, the design is completed before the construction works start but, practically, there is some level of overlap. The leader of the team of consultants undertakes the supervision duties.

This approach features the division of responsibilities between the designer and the constructor. Apart from being time-consuming – the design finishing before the construction starts – the client is not able to take full advantage of the contractor's capabilities and experience for use in the design development. In other words, there is virtually no input into design by the contractor. This may lead to problems with buildability, that is, the ease with which the designed facility is constructed. Although the client has a relatively high level of certainty of construction cost, and does not make a commitment to the whole works at an early stage, changes in design during construction may potentially cause an increase in cost and may result in adversarialism.

Direct labour

In this case the 'contractor' undertaking the works consists of employees of the client's organisation. In developing countries they are usually employees of the government. This organisation may also have staff with sufficient expertise to undertake the design of the works. If there is no design expertise a consultant is appointed to undertake the design, but usually not to supervise the works, on the basis of a fee contract.

This type of organisation is suitable for repetitive and intermittent construction works or where no specialist knowledge would be needed, for example minor reconstruction works on roads and bridges. It may be combined with other systems, particularly if work has been let in small packages.

Management contracting

The management system of procuring construction works has often been used for large works projects that require specialist services of several different types from trades contractors. The client appoints designers and a separate management contractor, who is paid for managing the construction works. The contractor is not expected to carry out any construction work. The works are normally packaged and let separately to works contractors of different disciplines, any number of packages can be in progress at any time, and so the contractor is expected to manage both the construction process and the package contractors. The design can also be the management contractor's responsibility.

This management procurement approach sees an early appointment of the management contractor to work alongside the design team and to develop a programme for the construction, design and

Figure 5.2. Management contracting

Figure 5.3. Design and build

tender documents. Thus, there is an opportunity for the contractor to bring skills to related to construction and buildability during the design development.

There are two common variations of the management system approach – management contracting, as shown in Figure 5.2, and construction management. In management contracting, the works package contractors are subcontracted to the management contractor, who manages and supervises them and arranges for their payment. In other words, the client is not privy to these agreements. In construction management, however, the package contractors are contracted directly to the client and the contractor only manages their works.

This type of organisation promotes teamwork and efficient use of resources through improved planning, but it is important to limit the overlap, in terms of supervision and monitoring, between the management contractor and the designer.

Design and build

This is a single-source approach in the sense that, generally, the eventual contractor takes on single-point responsibility to develop the entire design and deliver the works according to the client's requirements; see Figure 5.3. The client may appoint an independent advisor to monitor quality and cost, but generally the client has only a single point of contact.

There are, however, variations in the design and build approach. The develop and construct approach is one, where the client has the design prepared to concept or scheme design stage and the contractor takes on the finishing of the design and then the construction. Also, there is the package deal, where the contractor provides an off-the-shelf building.

The design and build approach is characterised by an overlap of design and construction, and any changes in design can have adverse cost implications on the works. The client also commits to both the design and the works at an early stage rather than the more gradual commitment associated with the traditional approach. Thus, any decision to abandon the works due to unforeseen conditions may have both cost and relationship problems. However, this approach offers minimal fragmentation and diverging interest, which can limit conflict to a manageable level.

The design and build approach is traditionally used for more straightforward buildings. But for developments with day-to-day operational requirements of very minimal disruption such as airport operations, this approach could cause a divorce between the contractor and the business operational requirements. In other words, the client may not be able to have the high degree of team integration and control that it needs, and the approach may therefore not be suitable.

This type of contract is preferred when the client organisation does not have the expertise or resources to undertake the works. The number of contractors capable of carrying out turnkey contracts will be limited, which will shorten the pre-qualification stage.

The evaluation of these bids is often more difficult because of the non-identical bids received. Bids are submitted in two parts, one containing the bid price and financial aspects and one containing the proposed design and technical specification. These bids have to make it clear exactly what the client will get for its money. The bids will initially be compared on technical merit and rated before financial evaluation. The financial evaluation on capital expenditure and running costs is then executed and rated. If further evaluation is required other parameters such as construction duration, process criteria and appropriate technology are assessed. Since no standard method of evaluation is available, the assessment is based upon a balance of objective and subjective criteria.

In this case the project manager must ensure that a detailed specification is produced and agreed by the client prior to tender. It should be noted that there is no subsequent opportunity for the client to change its requirements. The project manager must also take responsibility for checking compliance with the specification before certifying payments.

Framework agreements

As clients recognise the value of long-term arrangements framework agreements are becoming increasingly popular. A framework in itself gives no work to the contractor and may be non-exclusive. It is a long-term commitment between the parties to enable clients to place contracts on pre-agreed terms, specifications, rates, prices and mark-up that are embedded in the framework to cover a certain type of work over a period of time or in a certain location, or both.

The contractor makes staff, designers and construction resources available to undertake these contract packages as they are awarded and ensures their completion within agreed standards and time-scales. This is explored in detail in Chapter 11.

Partnering and alliances

The Reading Construction Forum, in *Trusting the Team* (Bennett and Jayes, 1995), defines partnering as follows:

> Partnering is a managerial approach used by two or more organisa-tions to achieve specific business objectives by maximising the effec-tiveness of each participant's resources. The approach is based on mutual objectives, an agreed method of problem resolution, and an active search for continuous measurable improvements.

This definition focuses on the key elements that feature promi-nently in partnering, irrespective of the form it takes, namely mutual objectives, an agreed method of problem resolution and continuous measurable improvements. Over the years the tradi-tional construction relationship has lacked any degree of objective alignment, and provides for no improvement in work processes. Parties enter the project focused on achieving their objectives and maximising their profit margins, with little or no regard for the impacts on others. This mindset leads to conflict, litigation and often a disastrous project. The characteristics of such a competitive environment include objectives which lack commonality and actu-ally conflict, and success coming at the expense of others (a win or lose mentality) and have a short-term focus.

Depending on the nature of the work being undertaken, part-nering can be categorised into two different groups:

- *Strategic alliances* (also known as term partnering) – for a period of time rather than a single project.
- *Project alliances* (also known as project-specific partnering) – arrangements for the duration of an individual project; the con-tract can be awarded competitively.

As well as these, there is also a variation of the latter which is commonly used within the public sector:

- *Post-award project-specific partnering* – here the contract is subject to the normal competitive process. As the name suggests, the partnering arrangement is entered into after the contract has been awarded. However, the intent to partner should be a crite-rion of the award process.

Partnering is fully described in Chapter 7. Alliance contracting is described in Chapter 8.

Payment mechanisms

Types of contract strategy are often classified by their payment system:

- *Price based* – fixed price and admeasurement. Prices or rates are submitted by the contractor in the tender. The client has no knowledge of the actual costs incurred.
- *Cost based* – cost-reimbursable and target cost. The actual costs incurred by the contractor are reimbursed, together with a fee for overheads and profit.

Fixed price

This is based on a single tendered price for the whole works. Payment may be staged at intervals of time or related to achieved milestones. The implications are that the complete, final design is available at tender or the contractor is to undertake the detailed design and that minimal changes or variations are expected. Under this system, the client places all or most of the risk with the contractor; it is therefore preferable that the risk is low or quantifiable, otherwise the client will pay a high risk premium. This leads to a high degree of certainty as to final cost, easy contract administration and little demand on the client's administrative resources.

This payment mechanism is usually teamed with 'turnkey' or 'design–build', the design being evolved by the contractor. It is therefore more likely to be suited to the contractor's own organisation and construction method, leading to savings in time and money.

Admeasurement

This is based on bills of quantities or schedules of rates, in which items of work are specified with quantities. Contractors tender unit rates or prices against each item. Payment is usually monthly and is derived from measuring quantities of completed work and valuing it at rates in the tender, or new rates negotiated from tendered rates if there has been a change.

Mechanisms are provided for adjusting both price and time in the likely event of change. This facility to introduce limited variation is frequently abused, as design may be only partially complete at tender. Extensive change and delay will generate claims, and the final price is invariably different from the tender total. The price will include an allowance for any financing required by the contractor and a risk contingency.

It should be used when design is complete but changes in quantity are expected, when little or no change to programme is envisaged and when the level of risk is low and quantifiable.

Cost-reimbursable

This is based on payment of the actual cost incurred by the contractor plus a fee specified for overheads and profit. The contractor's cost accounts are open to audit by the client (open book accounting). Payments may be monthly in advance, in arrears or from an imprest account.

Cost-reimbursable contracts are normally used when the client wishes to employ a contractor at an early stage of a project definition and design, or when there are major risks associated with the contract and/or access is severely restricted or likely to be disrupted by uncontrollable events. It is also useful when the work is innovative and productivities are unknown. There is little contractual incentive for the contractor to perform, and the final price will depend upon the extent to which risks materialise and on the efficiency of the contractor. Here, there must be positive cost control from all parties. The client carries the risk and will therefore require to participate in contract management.

Target cost

This is based on the setting of a probable (or target) cost for the work. The target cost will subsequently be adjusted for major changes in the work and cost inflation. The contractor's actual costs are monitored and reimbursed as in a cost-reimbursable contract. Any difference between actual cost and target cost is shared in a specified way between the client and contractor. There is a specified fee for overheads and profit. Difficulties can arise in agreeing when the target cost should be adjusted. Tender competition is improved. There can be problems in the timing of the supply of design information if the consultant does not have the same incentive as the contractor.

Target cost contracts have been successful in achieving a high degree of collaboration between the parties. They are most suitable for high-risk contracts where the work content is well defined, when the work is technically complex, when design input from the contractor or design flexibility is desirable and when the client wishes there to be some degree of training or development of local, skilled labour. Incentive arrangements are described in detail in Chapter 6.

Conditions of contract

Model forms of General Conditions of Contract have evolved in the UK to satisfy the different requirements and peculiarities of the various sections of the construction industry – building, civil engineering, mechanical and electrical plant erection and process plant. They are published by the relevant professional institutions and have gained international repute; their use and well-documented knowledge of the associated case law facilitate tendering and reduce the risk of misinterpretation. Each has been revised, and successive editions have incorporated minor changes to overcome problems encountered in their use and to reflect changes in both construction methods and contract management. The wording is precise and should not be changed without very good cause; amendment, where necessary, should preferably be by additional clauses.

The General Conditions will prove suitable for the majority of conventional construction contracts, but the client must satisfy itself that they are relevant to the particular job. The most common causes of difficulty and/or conflict arising from their use are a lack of sufficient or precise information at the time of tender and the introduction of a large amount of variation and change during the course of the contract. In the first case, it may be preferable to use a form of General Conditions but to delay tendering until the work is better defined. Otherwise, and when an excessive amount of change is anticipated and is unavoidable, a cost-reimbursable contract should be considered; special Conditions of Contract must then be formulated.

The various model forms of General Conditions of Contract are briefly mentioned below.

The Institution of Civil Engineers Conditions of Contract

The ICE Conditions of Contract are in their seventh edition (1999). This is an admeasurement contract using a bill of quantities prepared in accordance with the prescribed standard method of measurement. It is applicable when the design of the permanent works is prepared by another organisation, such as a consulting engineer, and where the design is substantially complete. Various clauses limit the effects of uncertainty and the relatively high commercial risk that arises from the diversity and nature of civil engineering works. The Engineer named in the contract is responsible for financial settlement. ICE Minor Works (2001) is used for low-risk work up to a value of £250 000, with a variety of types of payment available.

The Institution of Civil Engineers Design and Construct Conditions of Contract (2001) are a fixe- price form of contract where the contractor undertakes the design to a specification provided by the client. There is no independent Engineer.

The Engineering and Construction (NEC) Contract Forms

The Engineering and Construction Contract Form, ECC (1995), has been designed to promote good project management and is suitable for use for a wide variety of projects both in the UK and abroad. The language used is plain, present-tense English for ease of use, although this has attracted criticism from legal circles. There are core clauses that form the basis of the contract, with six options relating to the differing payment mechanisms available. The six options are: A, activity schedule with lump sum; B, remeasurement with bill of quantities; C, target set by sums but paid by actual cost; D, target set by rates and quantities but paid by actual cost; E, cost-reimbursable; and F, management contract paid by fee and actual cost.

There is a short form of the contract (for low-value, low-risk work), a term contract for term maintenance work, a professional services contract and an adjudicator's contract.

The Institution of Chemical Engineers Contract Forms

The Institution of Chemical Engineers publishes two popular forms of Model Conditions for process plants. The Red Book (1995) pays the contractor on a fixed-price basis and is used for design and construct projects where the client provides a performance specification and the contractor carries out the detailed design. It has a focus on construction completion, takeover and performance tests to ensure that the works are acceptable and perform to the pre-agreed standards. Failing these, damages are applied. In civil engineering, it is most applicable to water and sewage treatment works. The Green Book (1992) pays the contractor on a cost-reimbursable basis and has been widely amended to incorporate incentive mechanisms. It has frequently been used for high-risk civil engineering work, such as tunnelling, because of its risk-sharing and payment mechanisms. A Minor Works form is also available.

Government contracts

The Government issues GC Works 1 Edition 3 (1990), which has a lump sum payment mechanism with client design; GC Works 2 Edition 2 (1990); and PSA/1 with Quantities (1994). Although issued by the government for its own use, they have been adopted by other organisations for building and civil engineering work.

Institution of Electrical Engineers/Institution of Mechanical Engineers contracts

General Conditions of Contract Form MF/1 relates to the provision and erection of manufacturing and process plant, which may extend to the design and off-site fabrication of vessels or equipment to be supplied under the contract, erection on site, and testing and commissioning of the installation. This contract limits the total amount of variation to 15% of the contract price; any greater variation requires the consent of the contractor. There is no prescribed standard form for the schedule of prices referred to in these conditions; it is normally brief and simple when compared with the bills prepared for civil engineering work. The pattern of payment to the contractor is also very different, due to the importance of the supply and testing of fabricated units.

JCT contracts

The JCT Standard Forms of Building Contract are widely used for building construction. JCT 80 is a designer-led contract, the specification is not contractual, and the bill of quantities is consequently an extensive and detailed list of the components of the work. The expectation of physical uncertainty is considerably less here than in civil engineering contracts and only those billed items affected by variation are remeasured. Measurement and valuation are the responsibility of a quantity surveyor named by the architect in the contract. MW 80 is a fixed-price contract for simple designer-led contracts of short duration. The Intermediate Form (ICF 84) is for use on relatively simple projects (duration less than 12 months) without complex service installations. It can be used with or without quantities. There is also a management contract (MC87), a standard form with contractor design (WCD 81) and a standard form of prime cost contract (PCC 92).

International contracts

The FIDIC Conditions of Contract are recommended for use where tenders are invited on an international basis. There is the Red edition, which comprises general conditions based on the ICE 5th Edition and Conditions of Particular Application which must be drafted in detail for each specific contract. There is also a Yellow form based on MF1 for electrical and mechanical works, and a White form for professional appointments. The World Bank uses these conditions extensively and has defined a series of amendments.

Summary

This chapter has detailed the role of contract strategy and described the variety of procurement routes that are available. It has also provided an overview of some of the more common standard forms of contract.

Bibliography

Bennett, J. and Jayes, S. *Trusting the Team: The Best Practice Guide to Partnering in Construction.* Centre for Strategic Studies in Construction, University of Reading, Reading, 1995.

Smith, N. J. *Managing Risk in Construction Projects*. Blackwell, Oxford, 1999.

Turner, J. R. *The Commercial Project Manager*. McGraw-Hill, London, 1995.

Uff, J. and Capper, P. *Construction Contract Policy*. Centre of Construction Law and Management, King's College, London, 1989.

CHAPTER SIX

Incentivisation in construction contracts

D. Bower and B. Joyce

Introduction

The Latham Report (Latham, 1994) identified the need for the improvement of project and contract strategies to ensure greater satisfaction of clients. It suggested the need for 'basic principles on which modern contracts can be based as opposed to endlessly refining existing conditions which will not solve adversarial problems.' The European Commission (1994) also identified, as a result of dissatisfaction of clients, the need for change.

Incentives are used in contracts for two main reasons. One is to align the objectives of separate parties involved in a contract agreement; the other is to motivate the other party's behaviour to achieve the objectives set by the client. A way to set these objectives and to maintain continuous improvement is to use the tool of benchmarking and identifying key performance indicators (KPIs). This issue is addressed first in this chapter, and then the chapter describes the methodology of incentives, how they can be measured and where they can be used.

Benchmarking

Lord Simon of Highbury (1998), Minister for Trade and Competitiveness in Europe, in a letter to Tom Brock of the Global Benchmarking Network, wrote, 'Benchmarking is an important tool in enhancing a company's competitiveness. It is only by comparing itself with the best, that a company can see where and how it needs to improve its performance.'

Benchmarking is about comparing and measuring your performance against others in key business activities, and then using lessons learned from the best to make targeted improvements. It involves answering two questions – who is better and why are they better? – with the aim of using this information to make changes that will lead to real improvements. The best performance achieved in practice is the benchmark.

Benchmarking is the process of continuously comparing and measuring against an organisation anywhere in the world in order to gain information that will help your organisation improve its performance and competitive position. What makes benchmarking different from other management techniques is the element of comparison, particularly with the external environment. However, benchmarking is more than simple comparison, it is structured, it is ongoing, it compares itself with best practice and its aim is organisational improvement through the establishment of achievable goals

Types of benchmarking

Internal

This is a comparison of internal operations such as one office (or project team) against another within a company. It gives a company an understanding of its own performance level by identifying best practice in certain areas and transferring this to others. The reasons that practice within a company may be different in various sections may be because of factors such as location, history or the company being subject to buy-outs and mergers.

Internal benchmarking prepares a company for the process of external benchmarking in that it provides the data that will be required and ensures that those who will carry it out understand the process.

Competitive

This is a comparison against companies operating within the same business sector for the product, service or function of interest. The advantage of using this type of benchmarking is that it is specific to the company and comparisons can easily be found on the same level. The disadvantage is that the best practice of a competitor is not necessarily good enough. It may therefore be beneficial to look outside one's own industry to seek standards towards which to strive.

Generic

This is a comparison of business functions regardless of the industry that they belong to. Purchasing and recruitment are two examples of such functions. The advantages of generic benchmarking are that it breaks down the conventional barriers to thinking and offers an opportunity for innovation. As well as this, it broadens the knowledge base and can offer creative and stimulating ideas. A drawback is the cost in terms of time and money that this process consumes.

The benefits of benchmarking

To stay competitive, leading organisations regularly compare their own products, services and business processes against the best from within or outside their industry – seeking to unearth and implement best practice from whatever source. Organisations worldwide have found that there are significant gains to be made from benchmarking their activities, and that the amount of time and effort involved is repaid many times over.

Benefits include the facts that benchmarking:

- significantly reduces waste, rework and duplication
- increases awareness of what you do and how well you are doing it
- provides process understanding, leading to a more effective management
- helps set credible targets
- identifies what to change and why
- removes blinkers and 'not invented here' attitudes
- provides external focus
- enables the organisation to learn from outside.

It achieves these benefits by providing:

- a disciplined, realistic approach to assessing and improving the performance to be expected in critical areas of business
- learning from other companies' experience, so avoiding 'reinventing the wheel'.

Motivation

Provided that staff are kept informed every step of the way, benchmarking can be a great motivator. It requires an understanding of all areas of the business, requiring communication with the 'sharp end'. These staff then feel that they are valued and have a direct input into the benchmarking exercise and hopefully subsequent improvements.

Benchmarking to set performance targets

A barrier to creation of effective incentive contracts is the difficulty of establishing aggressive yet achievable targets that spur the contractor to higher levels of performance.

The use of an independent benchmarking consultant is one method used to establish credible yet challenging targets. A consultant maintains an extensive database of similar projects within the same industry to refer to. Using this information and results of similar projects conducted recently by the owner, the consultant can develop an estimate of what would constitute average contractor performance along with world-class contractor performance in the areas of cost, time, safety and other such identifiable quantities. These targets then become the basis for an incentivised contract.

A potential variation of the benchmarking process for incentive targets is to focus on functions that are specific to engineering and construction. While it is difficult to find a sufficiently large sample of similar projects that set end-of-project targets for cost and schedule, it should be easier to identify world-class performance for discrete tasks in order that improvement tasks may be set at a lower level. An example could be linking the amount of time required to bid, evaluate, and place an order for bulk materials to a financial incentive. Benchmarking can be an effective tool for establishing incentive targets when comparable projects or functions of projects are available to review.

Key performance indicators (KPIs)

A KPI is the measure of performance associated with an activity or process critical to the success of an organisation. The information provided by a KPI can be used to determine how an organisation compares with the benchmark, and is therefore a key component in an organisation's move towards best practice.

Companies and project teams need to objectively compare and benchmark their practices and performance, so that they can identify areas of improvement and thereby implement changes that lead to performance improvements. The purpose of a KPI is to provide an objective performance measure in a key activity associated with a company or project, which can be used (if the appropriate data exist) to compare and benchmark against the range of performance currently being achieved across other projects, companies or the rest of the industry.

Today, the definition of success is measured in terms of primary and secondary factors such as:

- *Primary factors*: on time; within cost; at the desired quality.
- *Secondary factors*: accepted by the customer; customer allows a company to use the customer's name as a reference.

It should be recognised that quality is defined by the customer, not by the contractor. The same holds true for project success. There must be customer acceptance; a project can be completed within time, within cost and within quality limits, and yet may not be accepted by the customer.

KPIs measure the quality of the process used to achieve the end results. KPIs are internal measures and can be reviewed on a periodic basis throughout the life cycle of a project.

The Construction Best Practice Programme (2002) states that its ten KPIs for 2000 were:

- *Project performance:*
  - client satisfaction – product
  - client satisfaction – service
  - defects
  - predictability – cost
  - predictability – time
  - construction cost
  - construction time.
- *Company performance:*
  - safety
  - profitability
  - productivity.

Incentivisation in construction contracts

Incentivisation is defined (HM Treasury, 1991) as

> A process by which a provider is motivated to achieve extra 'value added' services over those specified originally and which are of material benefit to the user. These should be assessable against pre-defined criteria. The process should benefit both parties.

Using an incentive should be considered in any contract but will be likely to be more relevant to contracts of sufficient scope and size to justify the investment in applying the technique; where the antici-pated benefits are assessed as being significant; and where, without incentivisation, improvements in performance and value for money would be unlikely to take place at a rate to match the business need.

A basic building block in the study of economic organisations is that individuals and firms do only what they perceive to be in their self-interests. Further, parties are assumed to be amoral in that manoeuvring, taking shortcuts, breaking agreements and taking any actions that offer personal gain are expected. This leads to useful predictions that often hold up even if the assumption is relaxed. Company incentive plans appear designed so that individual employees find it in their best interests to advance the company's goals, in spite of the fact that many employees would not stray the instant the incentives were removed. It is analytically useful to assume that the parties to the construction contract will focus purely on advancing their own interests. This leads more quickly to effec-tive contracting strategies.

This assumption does not eliminate seemingly unselfish behav-iour. This model is not necessarily made worthless by agreements being honoured, one's word being kept and more being done than is required by the contract. The notion of self-interests is broad, and when loss of reputation results in loss of business, this can be included within this area. A seller that values repeat business with satisfied clients has a long-term incentive not to take short-term gains at the expense of the customer and to the detriment of the seller/customer relationship. As an example, if the reputation of the company is at stake, a contractor may accept responsibility for faults outside of its control in order to maintain this reputation and so not to jeopardise future work with the owner.

The assumption of such self-interested behaviour relates to the conduct both of the owner and of the contractor because of their conflicting objectives. The owner will typically desire to obtain maximum quality, functionality and capacity at minimum cost. The contractor must achieve financial targets that will be greater the smaller the sum spent on resources to meet the scope of work set by the owner and yet still develop a satisfied client. These objectives are naturally in conflict.

Requirements of incentivisation in construction contracts

From a survey of European construction companies (ECI, 1999), it was found that some participants believed that all construction contracts must contain an element of incentive and of performance, otherwise no work would be undertaken. Other participants believed that it could be useful to subdivide the incentive contract type into two main categories: incentive contracts and performance contracts. The two categories of contract were defined as part of the research in the following ways:

- *Incentive-based contract:* a form of contract adopted by the client, which utilises 'positive incentives', typically monetary inducements to influence a service provider's performance, to the mutual advantage of both parties. Typical performance criteria to which incentives may be applied include project cost, schedule, safety and quality.
- *Performance-based contract:* a form of contract which formally specifies stringent performance criteria for the contracted service or facility, above and beyond generally accepted levels of competence or fitness for purpose. The fulfilment of defined performance criteria triggers full and final payment, while failure to meet defined performance criteria is typically associated with appropriate damages or penalty clauses.

Incentivisation of a contract requires:

- A clear and precise objective of what is to be achieved, both in delivering the service and value for money.
- A full understanding of the market and of the suppliers within it.
- Effective contract management by both parties and commitment to the incentives.
- An evaluation of the potential benefits at the procurement planning stage.
- The creation, by buyers, of the right culture of incentivisation by sharing their contract strategy with suppliers. Suppliers must see the process as a positive one, be willing to collaborate on the performance of the contract and to share information in order to effect continuous improvement.
- Effective pre-planning of payments (i.e. there is adequate provision to meet all potential incentive payments).

Incentivisation creates a more proactive, cooperative relationship between the parties of a contract, but this will be realised more readily only when the contract is of sufficient length to make the advantage apparent, assuming trust still exists between the parties.

The traditional view of incentives is that without an incentive mechanism system to monitor performance an organisation, and the people in it, will inevitably become slack and inefficient. Therefore the time and cost to completion of a project will always be greater if there is no real objective measurement of performance and control. The modern behaviouralistic theories of management and motivation may query this view. However, organisations and the people in them differ widely and it is essentially true that some people will always seek an easy life and, if they can get away with it, will settle down to a low, untrying performance level. Without effective incentive mechanisms, other people will not know they are not performing satisfactorily and will thus not apply the additional effort they could and would apply if they were aware they were not performing adequately.

Moral hazard

Owner and contractor objectives are not exactly aligned, and it is hard for the owner to calculate the quality of the contractor's personnel. Opportunistic behaviour by the contractor is possible, and the onus is placed on the owner to install opposing mechanisms which make such behaviour less likely.

The potential for such self-interested behaviour after contract execution is referred to as moral hazard, a term originally generated in the study of the insurance industry. If the contractor (agent) is acting on behalf of the owner (principal) and has interests that differ from the owner's, and if some aspects of the contractor's performance are not completely specified or observable, then the possibility of self-interested behaviour by the contractor at the expense of the owner exists. The nature of the capital-project process, where large investments are committed in advance of important performance by others, intensifies the moral-hazard potential. This can be more significant when the contractor stands to lose payment for some of the costs and not only the profit element is at risk.

Benefits of incentives

Incentivisation is a recognised and accepted contractual process used to achieve desired enhancements in performance over and above baseline requirements specified in the contract. Incentivised outcomes should be:

- to the benefit of all partners
- tangible and realistic
- measurable against an identified and agreed contractual baseline
- auditable
- such as to genuinely motivate all partners to take positive action to achieve improvement.

Some benefits that can be delivered by incentivisation in addition to those inherent in the base contract include:

- lower cost, faster or more timely delivery of service with no compromise on quality
- full understanding of the relationship cost, the quality of service delivery and the ability to deal more effectively with changes during the contract
- increased service levels
- greater price stability
- enhanced achievement of the desired outcome
- better utilisation of services
- improved management information
- improved management, control and monitoring of contract deliverables.

As shown in the ECI (1999) report, incentives, as well as having strengths, also have weaknesses:

- implementation is difficult – it requires a new approach
- resistance of construction culture to new approaches
- difficulty in designing tactical implementation to match intended strategic objectives.

The incentive of maintaining a reputation

Reputations can play a key role in ensuring contract compliance. Consider the situation of an owner who unfairly assesses a contractor in order to avoid paying incentive bonuses. This can be avoided by a system of quality enforcement mechanisms. When a buyer receives a bad unwarranted product, it is money lost. But if the buyer is convinced that the seller will suffer a substantial loss in the case of a defective product, the buyer is more likely to go ahead with the purchase. If cheated, the buyer and other sequentially informed buyers can impose on the seller a penalty by abstaining from purchasing the product in the future and by influencing others to do the same. The harder it is for the buyer to estimate product quality, the more important reputation is to the purchase decision, and the more likely the seller is to invest in reputation-enhancing activities such as advertising and customer service. Reputations and brand names not only indicate quality, they give the buyer a means of retaliation if the quality does not meet expectations. In effect, reputation can serve as a bond to ensure performance.

Similarly, concern over reputation should reduce the likelihood of a contractor neglecting areas not covered by incentives. Project results that damage the regard in which a contractor is held by the owner and others with whom the owner is in contact will result in a reduction in future work. In effect, a penalty results from bad performance although not directly from the contract.

The ratchet effect

Often it is difficult to determine appropriate performance levels on which to base incentives. From the owner's standpoint, each project may be unique, with variables such as scope, schedule, constraints and technical difficulty. The owner may wish to associate average compensation with average performance, and to tie maximum com -pensation to aggressive but achievable targets. What constitutes acceptable performance, and what constitutes outstanding performance, is difficult to determine.

There are basically two options that the owner can choose from to solve this problem. Performance levels can be based on comparisons with other contractors on similar projects, using the owner's previous experience, benchmarking, KPIs or other such information sources, alternatively, performance can be based on this contractor's previous experience on similar work. In long-term relationships or cases of repeat contracts, past performance offers the most reliable indication of performance.

Cost incentives

The motivational assumptions embodied in the use of cost incentives owe much to the principles of 'expectancy theory', a branch of learning and motivation theory which emphasises the use of rewards and punishments as incentives and reinforcers to influence behaviour. Another, more familiar way of referring to it is as the 'carrot and stick' approach. This approach is designed with the intention of 'shaping' behaviour through the selective application of rewards and punishments that encourage certain actions and discourage others. The effects of such processes of positive and negative reinforcement are potentially very powerful.

The essential elements of a cost incentive contract are:

- A target cost, which should be the best estimate mutually agreed by both contracting parties of what the costs will be when the work is done.
- A target fee, which is the amount of profit payable if the actual costs equal the target cost.
- The share formula, which describes the way in which any differences between the actual cost and the target cost are to be distributed between the contracting parties.

Fixed-price incentive contracts

A fixed-price incentive contract is a fixed-price contract that provides for adjusting profit and establishing the final contract price by application of a formula based on the relationship of the total final negotiated cost to the total target cost. The final price is subject to a price ceiling, negotiated at the outset.

The essential elements of fixed-price incentive mechanisms are the target cost, the target fee (which is the sum payable for overheads and profit if the actual costs equal the target cost) and the share formula (which determines how any under- or overrun of actual costs against the target will be shared). In addition, a range of mechanisms may be included to indicate the degree of confidence in the target cost and to impose limits on the price or fee.

A share formula is the most common way of inserting financial incentives into fixed-price contracts. The share formula gives the ratio in which the difference between the actual cost and the target cost is shared between the client and the contractor(s). The formula can be constant for actual costs above or below the target cost or it may be in the form of different sharing arrangements above and below the target cost. Typically the formula is expressed as a ratio such as 50/50 or 75/25, which shows the percentages of the saving or excess to be carried by the client and contractor respectively.

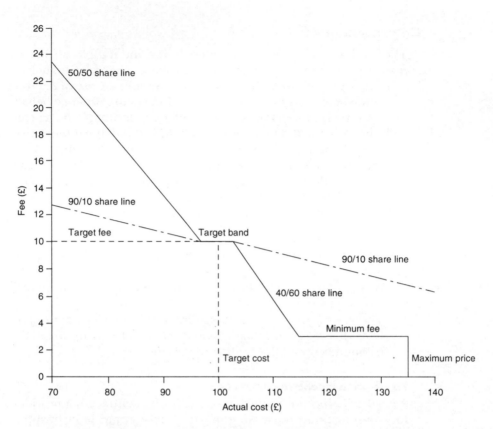

Figure 6.1. Illustration of target characteristics. (Note: this diagram does not necessarily represent a practicable system of incentives, but simply illustrates the meaning of terms.) Source: Perry and Thompson (1985)

It can be seen from Figure 6.1 that with a share formula of the order of 50/50 the effect on the profit margin of the contractor can be dramatic if there is a considerable variation between the actual cost and the target cost. Therefore this scale of share formula should only be used if there is a low risk that the actual cost will differ greatly from the target cost. If this risk is high, then the rate of change can be reduced by implementing a share formula in the order of 90/10, as shown by the dot–dashed line in Figure 6.1.

If appropriate, the sharing arrangement may be limited on the overrun side of the target cost by the use of a guaranteed maximum price. If the costs rise to that point the contract effectively reverts to a fixed-price contract. In other words, the share formula becomes 0/100.

A fixed-price incentive contract is appropriate when:

- a firm-fixed-price contract is not suitable
- the nature of the supplies or services being acquired and other circumstances of the acquisition are such that the contractor's assumption of a degree of cost responsibility will provide a positive profit incentive for effective cost control and performance; and
- if the contract also includes incentives on technical performance and/or delivery, the performance requirements provide a reasonable opportunity for the incentives to have a meaningful impact on the contractor's management of the work.

Cost-reimbursable incentive contracts

Cost-reimbursable incentive contracts include cost-plus-incentive-fee contracts and cost-plus-award-fee contracts. A cost-plus-incentive-fee contract is a cost-reimbursement contract that includes provision for the initially negotiated price to be altered later by a formula based on the relationship of total allowable costs to total target costs. A cost-plus-award-fee contract provides for a fee consisting of a base sum agreed and fixed at inception of the contract and an award sum that the contractor may earn in whole or in part during employment and that is sufficient to provide motivation for excellence in areas such as cost, time and quality.

In a cost-reimbursable contract it is usual for the contractor's off-site costs (including overheads and profit) to be paid on a fee basis, sometimes with profit the subject of a separate fee. The fee may be assessed as a percentage of actual costs or as a lump sum. A lump-sum fee constitutes the greater incentive to the contractor and is to be recommended. If a percentage fee is used, the contractor may wish a lower limit (in the form of a lump sum) to be imposed should actual costs fall below a specified minimum figure. If a fixed fee is used, the contractor may wish for renegotiation based on a maximum figure for the value of variations and delay.

When the contract is completed, the contractor submits a statement of costs incurred in the performance of the contract. The costs are audited to determine allowability and questionable charges are removed. This determines the negotiated cost. The negotiated cost is then subtracted from the target cost. This number is then multiplied by the sharing ratio. If the number is positive, it is added to the target profit. If it is negative, it is subtracted. The new number, the final profit, is then added to the negotiated cost to determine the final price. The final price never exceeds the price ceiling.

An alternative to the share formula as a means of introducing incentives is to apply a sliding-scale fee to the actual costs and it should be noted that this is simply an alternative way of describing the incentive and is not fundamentally different from a share formula. Some clients and engineers have considered that a weighted reduction in fee, on certain contracts, was advisable to ensure that the contractor avoided a 'loss of capital' situation. Here, a minimum fee excluding profit is fixed and the contractor is paid actual costs plus this minimum fee.

Are cost incentives a success?

Cost incentives may well have a significant influence on productivity but it is difficult, if not impossible, to identify and measure their degree of success independently of the other variables. Unfortunately, however, it is the accuracy of assessment which would indicate the validity of the scheme and its suitability for the job to which it is applied.

In basing pay on performance, however, the client also transfers risk to the contractor. Contractors may charge a premium to bear this risk, so the client's challenge in designing incentives becomes one of balancing the gain in contractor performance against the cost of shifting risk to the contractor.

Implementation of incentive mechanisms in construction contracts can meet with many difficulties. To maximise the potential benefits of such mechanisms, they should be used by companies with management expertise in the techniques. The variables associated with design, constructional method, technical difficulty, project type and work location should be reflected in the relevant targets. The chance of success associated with the incentive mechanisms will be greatly reduced if these criteria are not complied with.

The concept of financial incentive schemes dates from the philosophy of 'rational economic man', reasoning that people only (or primarily) work for money. However, theories by Maslow, Herzberg and others have shown that, above a certain level, money ceases to be a motivator. It is suggested, therefore, that in addition to financial incentives, other motivational methods be considered simultaneously for increasing employee productivity.

Famously, in the UK, the rail privatisation financial structure gave Railtrack little incentive to maintain or upgrade the infrastructure. Most of its income came from train operators, and over 90% of that was fixed, so there was no incentive to invest in expansion to provide more slots. This meant that as passenger demand grew, there were no more trains to carry the passengers. This scenario could have been avoided by using a financial incentive to encourage development. Additionally, penalties imposed by the government contributed eventually to the downfall of Railtrack (French-Thornton, 2001).

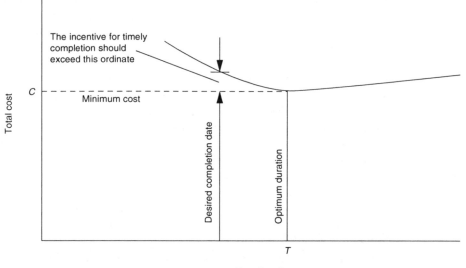

Figure 6.2. The use of incentives to achieve early completion. Source: Perry and Thompson (1985)

Time incentives

A form of time target exists in the majority of construction contracts as a liquidated damages clause, but it is difficult to enforce this when the work has been subject to variation. Where time is of great importance, direct penalty or bonus payments can be introduced into any form of contract to offer either positive or negative incentive, and these are often linked to key events in the programme. Figure 6.2 shows a typical time-related incentive mechanism (Perry and Thompson, 1985).

When the client desires completion in a period less than the optimum contract duration, this implies the provision of additional resources by the contractor. The natural inclination of a contractor on a fixed-price contract is to keep resources at the lowest level commensurate with the minimum-cost/optimum-duration condition. This suggests that the magnitude of the incentive should be considerably greater than the cost of providing the additional resources necessary to achieve the desired completion date. The application of such incentives in admeasurement contracts is sometimes confounded by contentious claims for extension of the contract period due to factors outside the contractor's control.

The advantage in using a cost-reimbursable contract, when time is the essence of the client's requirements, arises from the need for the client and contractor to be jointly involved in the early planning of the work. The client can then assure itself that the resources employed are adequate and exercise greater control over the flow of information to the contractor, thus minimising the chance of disruption. One interesting approach is a rolling bonus offered to a contractor to meet a sequence of time targets. The bonus increases in value, accumulating an increment as each target is reached: if a target is not achieved the bonus offered for the next target is reduced to the original minimum value. When the contract is awarded, the total number of targets is agreed and those in the first few months defined: the remainder are agreed as the job proceeds. The total bonus can amount to 10 or 15% of the total value of the contract.

Quality incentives

Bonus payments are made against the achievement of specified quality standards, such as the quantities of rework performed, or lack of defects due to poor workmanship. The achievement of quality standards may be measured by the number of hours lost over a given period due to time spent rectifying faults. Incentives may be similarly calculated on a sliding scale, increasing with a decrease in time lost.

It is generally possible to set targets based on quality only when the contract includes an operating plant. Performance incentives are therefore more common in the process plant industry, especially when the contractor is also responsible for the design.

Safety incentives

Bonus payments are made against the achievement of a specified safety record, typically measured by the frequency of reported incidents or the associated lost time in working hours. Incentives may be non-financial, taking the form of awards, prizes or gifts to worthy causes, such as a charity specified by the workforce.

Incremental or end-of-project payments

Incentive payments can be based on regular assessments that occur throughout the project or on end-of-project results, each of which carries its own advantages and disadvantages.

The benefit of continuous assessment and incentive payment is that it maintains more uniform incentive pressure, provides more timely performance reinforcement and prevents poor performance in one period from eliminating motivating incentive potential in other periods. Arguments in favour of end-of-project assessment and award are that end results are what are most important to the owner; that maximum incentive pressure needs to be applied in the final, critical periods of a project; and that incremental evaluations require excessive administrative time.

An incentivisation process

The incentivisation process described here has been developed on the basis of HM Treasury CUP Guidance No. 58 (HM Treasury, 1991), as it was found to be the clearest example of an incentivisation process and the most detailed. The purpose of the incentivisation process is to give an idea of the stages of elimination required in deciding whether and how to install incentive mechanisms into a construction contract. It provides the questions that need to be answered at each stage in the process and comments on the problems encountered when answering them.

When considering the questions to be answered regarding the type of incentive mechanism to use once the incentive criteria have been decided, the above cost, time and combined cost/time models should be considered.

Stage 1: initial planning
Step 1: establish justification of incentives
Questions to be answered. Does the size of the project justify the use of incentive mechanisms? Is there enough time available to create the incentive mechanisms? Can trust exist between the client and the contractor?

Comments. The benefits of incentivisation should be such that the use of incentives can potentially greatly affect the success of the project. If not, then the use of a standard contract will mean that the project has a greater chance of success, owing to the investment of time and capital in the administrative procedure of creating incentive mechanisms. The scope, size and duration of the project must justify the investment of additional resources. Because of the length of time required to create the respective incentive mechanisms, if a project is urgent then the amount of time available in which to design incentive mechanisms may not be sufficient.

The aspect of trust between the contractor and the client must be taken into account when deciding whether to incentivise a contract. If the two parties have never worked together before, the client should take more care when deciding whether to use an incentive mechanism or not than if a partnering relationship has existed between them for previous projects.

Step 2: define objectives

Questions to be answered. Does the specification contain output-based deliverables that can be identified? Are the objectives and the deliverables clear?

Comments. If no clear output-based deliverables can be identified there is no basis for an incentive mechanism. Likewise, any vagueness about the objectives will mean that incentive targets will be impossible to establish and therefore no incentive is possible.

Step 3: verify market stability

Questions to be answered. Is the market stable? Can prices be verified?

Comments. Incentives are only likely to work in a stable market, owing to their sensitivity to variations in the targets. For this reason it is necessary to verify the contractor's capabilities and knowledge acquired from previous projects and also to verify that prices can be verified as reasonable from benchmarking or KPI techniques. If the market is deemed to be unstable or the contractors available have limited experience in the relevant field, then incentive mechanisms should not be considered.

Stage 2: setting of baseline data

Step 1: establish direct costs and overheads

Question to be answered. Can stretched performance and cost targets be identified, quantified and set?

Comments. If incentivisation is still being considered at this stage, the baseline data for the costs of the project must be set. The client must establish the direct costs and overheads and the make-up of the contractor's costs. Once these costs are established, open book accounting can be used so that the contractor accepts the basis of the costs and to ensure that the understanding of any variation to the costs or schedule is fully comprehended.

Step 2: final evaluation of the need for an incentive

Question to be answered. Is an incentive necessary?

Comments. The final question to answer, when deciding whether to use an incentive mechanism in a contract or not, is whether the improvement of the chosen critical aspect(s) is possible to achieve without incentivising the contract.

Stage 3: measurement of benefits

Step 1: establish expected benefits to the client

Question to be answered. Can the benefits be expressed in terms of improvements that the contractor can deliver?

Comments. If the benefits are not measurable or quantifiable there is no means by which to set a target. This restrains the use of incentive mechanisms to such areas as cost, time, quality and performance by means of finite, measurable outcomes. Factors such as public satisfaction based on the number of complaints during the construction period cannot be used. This is because of the fact that not all members of public will know how to make a complaint, and if the process was made clear then more complaints would be made, apparently indicating poor performance rather than good communication techniques.

Step 2: express benefits in terms of improvements to the contractor

Questions to be answered. Will the contractor benefit from the contract? Are the benefits acceptable to the contractor?

Comments. The fact that the contractor carries a larger amount of risk if an incentive mechanism is included in a contract means that the benefits associated with successful target achievement should outweigh this risk. If not, the contractor may refuse to work under such conditions. In this case performance targets or the profit associated with reaching such targets must be altered to the satisfaction of the contractor.

Stage 4: setting objectives

Step 1: define framework

Questions to be answered. How are the savings/losses shared? How is the cost measured? What are the targets?

Comments. At this stage, the specification of the target details must be finalised. Decisions must be made according to the merits of the respective contract situation.

Step 2: define rules for scheme administration

Questions to be answered. Are clauses included for variations? Is inflation and cost adjustment allowed for in recalculating the target costs?

Comments. If targets are not updated, the initial purpose of the incentive is lost. The contractor must not be penalised for faults in the work that are not the contractor's fault. It will be impossible to make these adjustments totally accurate, as all prices will fluctuate slightly and the amount of administration involved will outweigh the benefit.

Stage 5: place contract

Questions to be answered. Does the contractor agree to the proposed incentives? Does the client have adequate resources available to manage the contract effectively?

Comments. Once all the details of the incentive mechanism have been finalised by the client, the contractor must agree to them. The client must then ensure that it has the required resources to manage the contract. If not, then extra personnel or other resources must be acquired before work commences.

Stage 6: evaluate contract

Questions to be answered. Was the contract a success? Did the incentive mechanisms improve performance? What problems were encountered? Were objectives aligned?

Comments. The success of the contract must be evaluated as part of an ongoing process so that problems can be quickly identified and resolved with as little consequence as possible. Once the contract is complete, a review of the project must be carried out in which the incentive mechanisms are objectively evaluated so that positive and negative factors can be identified and improvements can be made in the future.

Summary

Before deciding to incentivise a contract, it is essential to identify whether the improvement in quality and/or cost reduction can be achieved without incentivising the contract.

Incentives can positively impact contractor performance, but the performance gain must be balanced against the costs of shifting related risk to the contractor. Clients can broaden their incentive options by taking actions that increase the level of contractor control over contractor results, thereby reducing contractor risk.

Assessment of the degree of success of incentive mechanisms is extremely difficult. It has been shown that on a site where a direct financial incentive has been applied, increases in output could equally well be attributed to operative training, job experience, good supervision, the implementation of management techniques, repetition of work tasks or the extent of mechanisation.

Engineering and construction contractors' performance affects owner profitability long after project completion. Factors such as instrument and valve placement for maintenance, durability of specified finishes, and provision for future utility tie-ins impact the client's operation costs for the life of the facility. However, typical contracts do not tie the contractor's payment to these factors. This concept has very limited applicability. A contractor's influence on long-term facility profitability is difficult to separate from factors outside the contractor's control, such as market conditions. Also, the delay of cash flow to contractors that would result from linking the fee to long-term facility performance appears a major obstacle. Effective long-term incentives based on the contractor's contribution to owner success appear to be an opportunity for further research.

The general principles upon which incentive mechanisms should be based include the need to ensure that risks and rewards are commensurably and fairly distributed among the parties concerned and that they are tailored to specific project objectives.

An important factor in designing incentive mechanisms is to try to align the separate schedules of designers and contractors with those of the client. In order for this to occur, the objectives of all parties involved, especially those of the client, must be clear from the outset.

Incentive contracts are an attempt to align the interests of the contractor with those of the owner by basing compensation, to some degree, on results that are important to the owner. In basing pay on performance, however, the owner also transfers risk to the contractor. The output of a contractor is typically a function of factors within its control (such as level of effort, quality of assigned personnel, and management attention) and outside its control (such as weather, supplier problems and the owner's technology). The performance variation resulting from outside factors introduces randomness to the contractor's output, and therefore to its income. Contractors may charge a premium to bear this risk, so the owner's challenge in designing incentives becomes one of balancing the gain in contractor performance against the added cost of risk-bearing.

Although incentives can play an important role in making a project a success, a project's success is determined by the sum of its parts and incentives are just a small part in the make-up of a project. The basic aim of including incentive mechanisms in contracts is to use the contractor's main objective, that of making a profit, to encourage the contractor to make more money by performing the contract efficiently and in line with the client's objectives.

Care must be taken in designing incentive mechanisms for areas that are a basis of incentive payment. The natural attitude of a contractor is to concentrate effort where the potential return is greatest, thus perhaps leaving areas of neglect. A project may be completed within cost, time and safety targets, meaning the bonus criteria have been fulfilled, and yet, for example, the contractor could have triggered an environmental incident that affects neighbours to the construction, causing long-term problems. Incentives have the potential to distract effort from certain areas.

The responsibility of structuring an effective incentive mechanism in a construction contract lies with the client. The basis for selecting and designing this mechanism is for the client to consider its own objectives and experience. Performance areas should be correlated with key result areas for the owner, and must sufficiently span the contractor's performance so that important aspects of performance are not neglected.

One principle of incentive pay is that the more control the contractor has over performance areas covered by incentives, the higher the effectiveness of the incentives in shaping the contractor's performance.

Bibliography

Benchmarking (2002). Website: www.benchmarking.co.uk.

Construction Best Practice Programme (2002). Website: www.cbpp.org.uk.

DuBrin, A. J. *Effective Business Psychology*, 3rd edition. Prentice-Hall, London, 1990.

ECI. *Study of the Effectiveness of Incentive Strategies.* ECI, Loughborough, 1999.

European Commission. *Strategies for the European Construction Sector. A report compiled for the European Commission by WS Atkins International Ltd.* Construction Europe, Sussex, 1994.

French-Thornton (2001). Website: www.noflash.french-thornton.co.uk/press/press_article_jan01.

HM Treasury. CUP Guidance No. 58. HM Treasury, London, 1991.

Howard, W. E. and Bell, L. C. *Innovative Strategies for Contractor Compensation.* Construction Industry Institute Research Report 114-11. Construction Industry Institute, Austin, TX, 1998.

Kopelman, R. E. Linking pay to performance is a proven management tool. *Personnel Administrator*, Oct. (1983), 68.

Kpizone (2002). Website: www.kpizone.com.

Latham, M. *Constructing the Team: Final Report of the Government/Industry Review of Procurement and Contractual Arrangements in the UK Construction Industry.* HMSO, London, 1994.

Locke, E. A. and Latham, G. P. *A Theory of Goal Setting and Task Performance.* Prentice-Hall, Englewood Cliffs, NJ, 1990.

Lord Simon of Highbury. Letter to T. Brock of the Global Benchmarking Network, 24 July. Department of Trade and Industry, London, 1998.

Perry, J. G. and Thompson, P. A. *Target and Cost-reimbursable Contracts.* Report R85. CIRIA, London, 1985.

CHAPTER SEVEN

Effective partnering

D. Bower

Introduction

As early as 1964, the Banwell Report (Banwell, 1964) noted that the UK construction industry was performing poorly compared with other countries. This state of affairs had not improved by the early 1990s, when the industry was perceived as being in decline. One of the contributors to this was the fragmentation of services, leading to a high proportion of contractual relationships and an increasing reliance by the industry on litigation procedures as disputes on site between clients, contractors and subcontractors were dragged into the courts. The increasing number of contractual disputes has had the knock-on effect of more and more projects being delayed, and the overall efficiency of the industry being reduced. There has also been the problem that the industry has lacked a focus on customer satisfaction – a trait that seems to be almost unique to the construction industry.

Since the Banwell Report, there have been a series of government and industry reports that have attempted to find solutions to this problem of fragmentation and efficiency. Partnering is the latest initiative to have come out of this. Partnering, which involves clients and contractors developing a closer, possibly longer-term working relationship, has been implemented as a way of increasing efficiency within the industry, as well as a way of sharing risk.

This chapter examines partnering from its foundation in the concept of total quality. It then goes on to outline its key features and the benefits that can be accrued from its adoption. Finally, a framework for best practice is described.

Total quality management

The concept and application of total quality management (TQM) has been successfully integrated into manufacturing and service industries worldwide. It entails a company-wide effort that involves everyone in the organisation in the effort to improve performance and customer satisfaction. It focuses on process improvement, customer and supplier involvement, teamwork, and training and education in an effort to achieve customer satisfaction, cost-effectiveness and defect-free work. Continuous improvement is attained through the application and integration of both human resources and quantitative methods. There are six basic concepts that are required for TQM:

1. A commitment by management to provide long-term top-to-bottom organisational support.
2. An unwavering focus on the customer, both internally and externally.
3. Effective involvement and utilisation of the entire workforce.
4. Continuous improvement of the business and production process.
5. Treating supplies as partners.
6. Establishing performance measures for the processes.

The concept of TQM has been embraced and worked successfully in the manufacturing and service industries for various reasons. Manufacturing-based quality programmes tend to be product oriented, focusing on changes that improve the manufacturer's completed product. Service quality programmes are more process and personnel oriented because process improvements that result in better service to customers are the main goal of service quality management programmes.

The construction industry incorporates elements of both manufacturing and service processes, which makes it difficult to standardise one format or the other for the entire industry. Another reason for the difficulties in using TQM in construction is that projects are mostly one-off and unique, meeting a specific timeframe, and so it is difficult to engender continuous improvement. The projects are also realised in uncontrolled environments and involve enormous resources over a relatively short span of time. In spite of the difficulty in achieving total quality, most construction organisations have achieved certification. This has been gained without regard to delays in project delivery, cost overruns and the frequent low quality of products.

This rush to certification can be explained by the fear of many construction organisations that without the standard they risk losing work. The certification has turned out to be a means of guaranteeing continuous work rather than providing quality to the client. An important aspect of the overall change brought about by a TQM approach is a changed relationship with suppliers. The traditional approach of construction, which typically organises projects by hierarchically linked parties (clients, consultants, general contractors, subcontractors, suppliers, etc.) possessing different skills and knowledge, results in complex and adversarial relationships, which affect performance.

It is therefore in the desire to deliver projects within budget, on time and to acceptable quality (which TQM has not been able to offer) that project partnering finds its roots. Increased international competitiveness, enhanced legal concerns, the introduction of new technologies, and the desired response time to delivery have also necessitated the need to change the traditional approach to project delivery and have resulted in the evolution of partnering.

Partnering

Partnering has been widely advocated for the industry in the UK to rectify the adversarial contractual relationships that have jeopardised the success of many projects (Latham, 1994; Baden Hellard, 1995; Construction Industry Board, 1997; Bennett and Jayes, 1998).

Features of partnering relationships have been seen in various industries for many years. The partnering style of relationships with contractors was a feature of some construction projects in Britain early in the Industrial Revolution (Barnes, 2000). As applied today, it originates in the philosophies of the Japanese-influenced automobile industry. The defence, aerospace and construction industries have followed. Its essence is *alignment* of values and working practices by all members of the supply chain in order to meet the customer's real needs and objectives, though this has been pursued with different degrees of success and sustaining it is a questionable objective (Green, 1999). Continuous improvement has been an important objective, with emphasis not only on cost but also on quality, lead time, customer service, and health and safety at work. Incentivising the partnering companies by sharing cost savings has been a feature of continuous improvement, performance -based partnering in many industries, but in construction this has often been less significant than the primary objective of avoiding disputes.

The idea of alignment is significantly at odds with traditional practice in many industries. Procurement in most of the public sector has historically been based on accepting the lowest-price bid. Much private construction also traditionally operated on this basis. It has led to conflicts about paying the actual costs of work, which revolve around risks and financial self-interest, between the various stakeholders – such as the clients, design team, consultants, main contractors, subcontractors and suppliers – throughout the construction process. As a consequence, the final cost of the project usually exceeds the contract price and the result is confrontation.

Partnering represents a philosophy of dispute prevention, conflict resolution and equitable risk allocation rather than a legalistic and confrontational approach. Partnering is an addition to other good project management practices, long-standing relationships, negotiated contracts, preferred supplier arrangements and other forms of team-based supply chain arrangements. Partnering is an addition to the techniques for dispute management. As shown by the Movement for Innovation projects, partnering can be applied to all types of contract arrangements and with any of the industry's model terms of construction contract, and it can extend to subcontractors.

The objective is to create a 'win–win' culture so that projects are completed successfully and so recover the confidence of clients in the industry. Partnering is a process to establish good relationships at all the interfaces between stakeholders, and their commitment to the job and each other. Partnering should create trust, teamwork and cooperation to give early warning of potential problems and establish effective authority to agree decisions on them. It is critical throughout a project to remove traditional barriers and perceptions of unfairness between the parties involved. By changing to a 'win–win' style the parties can reap benefits of cost saving, profit sharing, quality enhancement and time management. Unifying all the parties into one team for a project, it also reduces transaction costs.

The intention of partnering is to change the commercial style in which contracts are managed. It demands a shift from concentration on 'hard issues', such as price and the scope of work, towards 'softer issues' that revolve around attitude, culture, commitment and capability. 'Successful partnerships manage the relationships, not just the deal' (Kanter, 1994). The culture and attitude of its participants have to be changed to develop a single social network. Ellison and Millar (1995) defined a four-level approach. From 'arms-length adversaries', the parties move to an environment of trust and communication, advancing to a 'partnering/integrated team' arrangement. The last step in this evolutionary chain is the most theoretical and difficult to achieve, a 'synergistic strategic partnership'. Trust is the essential ingredient.

Partnering has been defined in a variety of ways:

> Partnering includes the concepts of teamwork between supplier and client, and of total continuous improvement. It requires openness between the parties, ready acceptance of new ideas, trust and perceived mutual benefit .... We are confident that partnering can bring significant benefit by improving quality and timeliness of completion whilst reducing costs.

(Sir Michael Latham, *Constructing the Team* (Latham, 1994), quoting the Chartered Institute of Purchasing and Supply.)

> Partnering is a long-term commitment between two or more organizations for the purpose of achieving specific business objectives by maximizing the effectiveness of each participant's resources. This requires changing traditional relationships to a shared culture without regard to organizational boundaries. The relationship is based upon trust, dedication to common goals, and an understanding of each other's individual expectations and values.

(Construction Industry Institute, 1991.)

The above definitions depict partnering as a generic term and emphasise that the relationship will cause all to seek win–win solutions, place value in long-term relationships and encourage trust and openness to be the norms, and that an environment for profit exists. It is also a view that neither partner should benefit from exploitation of the other, innovation is encouraged, and each partner is aware of the other's needs, concerns and objectives and is interested in helping its partner achieve them. It creates a team environment to accomplish a set of goals in much the same way that a sports team works together to achieve its goals. But perhaps the definition that provides explicit meaning, which is adopted for this chapter, is that by the Reading Construction Forum, in *Trusting the Team* (Bennett and Jayes, 1995):

> Partnering is a managerial approach used by two or more organisations to achieve specific business objectives by maximising the effectiveness of each participant's resources. The approach is based on mutual objectives, an agreed method of problem resolution, and an active search for continuous measurable improvements.

This definition focuses on the key elements that feature prominently in partnering, irrespective of the form it takes, namely mutual objectives, an agreed method of problem resolution and continuous measurable improvements. Over the years the traditional construction relationship has lacked any degree of objective alignment, and provides for no improvement in work processes. Parties enter the project focused on achieving their objectives and maximising their profit margins, with little or no regard for the impacts on others. This mindset leads to conflict, litigation and often a disastrous project. The characteristics of such a competitive environment includes objectives which lack commonality and actually conflict, success coming at the expense of others (a win or lose mentality), and have a short-term focus.

The partnering process

The key to partnering is that it starts at the outset of a project (Matthews, 1996). The process is formally established by workshop (or 'kick-off') sessions between the partnering members so that everyone has a clear understanding of what the process is and agrees to use it. As in any collaborative venture, all parties have to get together pre-construction and invest time into agreeing and under-standing the objectives, form and operation of a partnering agreement (Wearne and Wright, 1998).

A competitive relationship is maintained by a coercive environment with little or no continuous improvement. Points of contact between organisations are mostly single, which does not encourage good interaction, and culminates in little trust, with no shared risk; this is primarily a defensive position. Thus a partnering relationship involves the essential elements of mutual objectiveness, problem resolution and continuous measured improvement. The Construction Industry Board (1997), in *Partnering in the Team*, details these as being:

- *Establishment of agreed and understood mutual objectives:* the objectives are agreed and committed to at the outset of the project, and kept under review through meetings and effective communications. They require long-term goals – sustained reasonable profitability rather than a quick killing. They benefit from 'open book' relationships (which also reduce the risk of corruption), treated with mutual confidentiality, resulting in the partners working for each others' success. It works best between businesses with similar cultures and styles.
- *Methodology for quick and cooperative problem resolution:* partnering sets up a systematic approach to problem resolution, seeking solutions rather than parties to blame; more and better discussion with less paperwork; more constructive correspondence, based on 'win–win' solutions; and equality of rights between parties. It requires mutual acceptance of the principle that adversarial attitudes waste time and money.
- *Culture of continuous, measured improvement:* partnering recommends that there should be specific quantified targets, measured progress and periodically reviewed performance. It allows that competition is not the only way to achieve best value for money but is customer focused, adding value, eliminating waste, and identifying and aiming for best practice.

It is important to note that partnering by itself does not produce any value for the client, but requires the full participation and effort

of all parties involved in achieving the desired goal. It must therefore not be misconstrued as:

- A new form of construction contract – it is a procedure for making relationships work better.
- An excuse for not working hard to get the best from suppliers and customers.
- A soft option.
- A quick fix for a weak business – strong players make each other stronger, weak ones destroy each other.
- Only about systems and methods – it is about people, enabling them to operate more effectively and efficiently.
- A panacea. Partnering will not prevent all problems in every contract. There may be some issues that must be litigated.
- Mandatory. Partnering is not a contractual requirement. It is a working relationship and if commitment is not present, it will not work.

Advantages of partnering

Partnering relationships offer advantages and opportunities specific to the individual members of the project team as well as the opportunities and advantages shared by each.

Benefits for client

Effective utilisation of personnel resources may be the most important benefit to the owner, in terms of both staffing requirements and
available expertise. The client may also benefit from increased flexibility and responsiveness in terms of added skills and resources available from other parties, from the presence of a diversity of talent not
usually found in a single company, which will improve on delivery,
and from reduced costs associated with contractor or consultant
selection, contract administration, mobilisation, and the learning
curve associated with beginning a project with a new contractor or
consultant.

Other benefits to the client will be the reduced dependence on
legal counsel, the development of a team for future projects and
more control over possible cost overruns.

Benefits for design team

Partnering provides the design team with the opportunity to refine
and develop new skills in a controlled and low-risk way. This occurs
because new methods or approaches may be required to meet
owner project requirements. Through partnering, the design team

will benefit from the involvement of contractors during budgeting, development of the team for future projects and optimal use of the design team's time.

Benefit for contractor

Although a partnering relationship will not make a specific guarantee of workload, partnering implies a clear intent to maintain an active functional organisation. The long-term, non-adversarial aspects of partnering mean that revenues may be more stable and the potential for the claims or litigation process is significantly reduced. The contractor may also benefit from increased opportunity for value-engineering involvement to provide value for money, faster decision-making processes, and more effective time and cost control.

Other benefits will include formation of teams for future projects, increased opportunity for financially successful projects, reduced dependence on legal counsel and the possibility of faster payments.

Benefits for the manufacturers and suppliers

As with the other team members, the benefits that manufacturers and suppliers stand to gain through partnering include approval of their product recommendation, a voice in the design intent, involvement in the coordination with other project trades and the possibility of repeat business. Other benefits are a better chance for quality in product installation and increased opportunity for financially successful projects.

Mutual benefits

Of all the potential benefits resulting from partnering relationships, perhaps the one that will have the most impact on the construction industry is improved project quality. An effective partnering agreement will improve project quality by replacing the potential adversarial atmosphere of a traditional owner–contractor–consultant relationship with an atmosphere that will foster a team approach to achieve a set of common goals.

Within this atmosphere of cooperation and mutual trust, the companies can jointly determine and evaluate approaches to designing, engineering and constructing the project. By becoming partners in the project, team members can work together to achieve the highest level of quality and safety. The close, team-working relationship between the parties can provide an environment that encourages finding new and better ways of doing business. An effective partnering relationship will encourage partners to evaluate technology for its applicability to quality improvement for the project.

Table 7.1. Different forms of partnering

Forms of partnering	Relationship duration	Basis of partner selection	Condition for use
Project	One-off	Competition/ negotiation	All projects. Best for high value
Strategic/full	Long-term	Competition/ negotiation	Where good business case, part of medium– long-term strategy
Post-award	One-off	Competition	Public projects, including series of small projects
Preselection	One-off/long-term	Negotiation	Any project. Advanced selection of contractors
Coordination agreement	One-off/long-term	Competition/ negotiation	Any project. Agreement overlaid on standard contract
Semi-project	One-off	Limited competition	All projects where scope of negotiation is limited

Source: Institution of Civil Engineering Surveyors (1997)

The partnering relationship also encourages the companies to identify major obstacles to the successful completion of the project and to develop preventive action plans to overcome those obstacles before they impact schedule or cost.

Forms of partnering

Partnering can be categorised into the following forms: project partnering, strategic/full partnering, post-award project partnering, preselection arrangement, coordination arrangement and semi-project partnering. These are shown in Table 7.1, and the key differences relate to relationship duration, basis of selection and the most appropriate conditions for application.

Post-award project specific partnering

This type of partnering is used for contracts that undergo the normal competitive processes but for which the intention to adopt a partnering approach throughout the project is declared during the tendering process. Here the concept of partnering is applied under the main contract for a particular project. The partnering application is detailed as part of the project contract document and both parties agree to overlay their formal contract with a partnering arrangement.

Workshops are held at the start of the project for the client team, the main contractor and any key subcontractors/suppliers to facilitate team building, clarify the aims and objectives of the parties, agree joint objectives for the project, develop processes and procedures for communications and problem resolution, and produce a partnering charter for the project. The workshops are also held at key stages of the project and at such times that are deemed necessary.

Generations of partnering

Partnering can take many different forms but the benefits that derive from these forms are achieved through relationships built up over many years. In the Reading Construction Forum report *The Seven Pillars of Partnering* (Bennet and Jayes, 1998), three distinct stages (known as generations) can be identified.

First-generation partnering

This type of partnering is formed when a construction business and its customers enter into a relationship to deliver a single project. First-generation partnering revolves around three key principles that apply to project teams. The first principle is that mutual objectives should be agreed to take into account the interests of all parties involved. Secondly, decision-making and problem solving should be done openly and in such a way as was initially agreed at the start of the project implementation. Lastly, targets should be aimed at continuous measurable improvements in performance from project to project. Decision-making encompasses problem resolution. The benefits of first-generation partnering include faster construction times, improvement in quality, less litigation, improved safety, better teamwork, more innovation and cost savings of 30%.

Second-generation partnering

This is a well-established type of partnering amongst leading firms and consists of partnering by a group of consultants and contractors who add a long-term strategic dimension to a series of projects for one customer. They jointly establish a strategic team that builds up the pillars of partnering. In this type of partnering, strategic membership, equity integration and benchmarking feedback encompass mutual objectives, decision-making and continuous improvement, respectively. It also involves establishing standards and procedures that embody best practice based on process engineering.

Third-generation partnering

This is spearheaded by construction firms that organise their business to provide continuity of workload. They do this by applying partnering throughout the whole supply chain to produce products designed for specific categories of customers. This level of partnering is mostly used in private finance initiatives where the private sector undertakes the work and leases the resulting facilities or service to public sector bodies.

At this level, firms use cooperation throughout their supply chains to build up efficient virtual organisations that respond to and shape rapidly changing markets. They harness new technologies to satisfy customers' expectations. In doing so they combine the efficiency that comes from standardised processes with the flexibility that comes from creativity and innovation. About 50% cost savings and 80% reduction of construction times have been suggested to be associated with this level of partnering.

Framework for best practice

Partnering can be used successfully if a framework that clearly identifies the critical issues that are necessary for its implementation is followed. These issues are highlighted below.

Management commitment

All organisations represented must have the complete support and commitment of upper management. This is essential, since all other members will look to this commitment to assess their own individual commitment. The commitment must be clearly visible to all project team members and, most importantly, should be sustained throughout the project duration.

Corporate culture

Managers educated in traditional business relations find partnering relationships threatening to their company. These managers' attitudes have been influenced by their company's corporate culture. A company's corporate culture is strongly influenced by what is important to the company. As a result of this inherent attitude, companies must strive to achieve effective 'internal partnering' horizontally between departments and vertically in the management structure.

As the business environment for the construction industry changes, a company's success may depend on recognising and adapting its culture to these changes. Managers must recognise changes in the business environment for their customers, suppliers and competitors as well as their own company, and managers must recognise how these changes will influence their corporate culture.

Team selection

Taking the time to identify the right partner is important to the success of a partnering relationship. It is also important to note that the inability of any one member to perform in the team would impact negatively on the success of the partnering arrangement. Several potential partners of a high quality with plenty of experience should be approached. A review of any potential partner's key staff skills and gaps should be carried out and their strengths and weaknesses assessed. The potential partner's management style, organisation and cultural differences must be analysed for the best fit.

Workshop

This is vital to any partnering project, no matter how often the partners have worked together before, or whether the partnering agreement is project specific or strategic. The workshop should be run by an independent facilitator and should be held at a neutral location.

The workshop agenda should include the following:

(a) An introduction by the clients' partnering champion and a sharing of views and expectations of the workshop.
(b) An overview of the partnering process given by the facilitator.
(c) Presentations identifying the usual causes of conflict, and discussions on avoidance.
(d) Problem-solving exercises using line issues, attempting to secure mutually advantageous solutions and establishing a project code of ethics. These activities should be undertaken by small teams, reporting back to the plenary group.
(e) News of improving performance/productivity.
(f) Identification of issues that may be barriers, problems or opportunities. These issues may relate to the partnering process, the project or both. These activities will use brainstorming techniques and encourage the formation of constructive ideas. They will be monitored by the facilitator.
(g) A summary of the issues, which may be allocated to teams to develop specific action plans, followed by a group discussion and finalisation of the action plans.
(h) Establishment of a dispute resolution procedure.

(i) Establishment of monitoring criteria and the monitoring procedure.

(j) Election of champions.

(k) Drawing up of a partnering charter, assisted by the facilitator.

(l) Workshop summing up by the partnering champions of both client and contractor.

Mutual objectives

The workshop will enable the partners to brainstorm each other's objectives, focusing on their end objectives as well as the intermediate ones. The determination of each other's criteria for the success of the project will help both parties to be aware of the individual expectations, and gives them adequate time to try to incorporate those objectives into the win–win objectives of the relationship. The objectives are usually captured in the form of a partnering charter, as shown in Figure 7.1.

Conflict resolution

Conflicting issues are common among parties with incompatible goals and objectives. Even with parties that have compatible goals, the problem cannot be wished away, because it is human centred. The impact of conflict resolution can be either productive or destructive and largely depends on the manner in which partners resolve problems. For enhancing cooperation and greater promise of long-term success, organisations are advised to adopt more productive resolution techniques such as joint problem solving, which is described as the collective decision to create alternatives for issues. A typical problem resolution flow chart is shown in Figure 7.2.

Continuous improvement

The partnering team must strive to achieve continuous improvement, since without it most of the benefits from it will be lost. A series of steps towards achieving continuous improvement have to be taken as follows:

(a) Clear goals have to be established.

(b) People have to be convinced that change and improvement are in their interest.

(c) Organisational barriers to change have to be eliminated.

(d) Open forums for exchange of views and debate have to be set up.

(e) A simple measurement system should be set up which immediately focuses attention on key issues.

Figure 7.1. Typical partnering charter, from Construction Industry Board (1997)

(f) Ideas should be systematically assessed to ensure that they can be developed into opportunities.

(g) Targets should be continually reviewed.

(h) Management interest and appreciation should be continually displayed.

(i) There should be a no-blame attitude.

(j) Recognition and reward must be maintained.

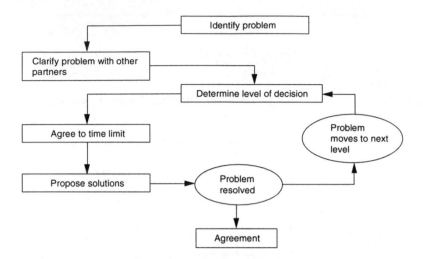

Figure 7.2. Problem resolution flow chart. Source: Bennett and Jayes (1995)

The elements essential to achieving improvement are the attitudes of the partnering teams, the knowledge base and information-sharing attitude of the partners, the techniques adopted or applied for the partnering arrangement, and the training given to 'individual boundaries' and 'think as a team'.

Potential barriers in implementing partnering

Partnering is based on relationships. When two firms negotiate a partnering agreement, they have a concept of fairness in sharing and in relative benefit. However, any relationship requiring mutual dependency and high levels of trust is bound to be confronted with certain barriers. These barriers must be known and expected so that they can be dealt with to achieve the objectives set out for the project. The barriers can be grouped as external and internal.

External barriers

The main external barrier to partnering, particularly in the public sector, is that of politics. When politicians decide to play politics with partnering, they can always misrepresent the notion of a mutual relationship as not protecting the public interest; this will lead to a large amount of pressure on public sector officials. For example, if public officials decide to sidestep the 'public sector business-as-usual' process of lengthy review, claims and possible litigation to achieve faster communication and response, they could be considered as not defending the public purse. Another barrier to partnering is the externally imposed law, permitting very strict notice provisions and aggressive exculpatory language in their contracts, which create different incentives for conflict resolution or prevention.

The last external barrier to partnering is when the reward system for middle management or project-level staff is determined externally and also includes disincentives to decision-making and taking risk. Partnering involves greater investment in decision-making and risk taking; therefore structures must be put in place to support and initiate that. If the system is such that there is a disincentive for taking decisions and risks then the likelihood of partnering succeeding will be decreased.

Internal barriers

The potential barriers that can affect the approach to partnering within an organisation are organisational structure, cultural attitudes and climate. Organisational structures have evolved within firms as a result of communication restraints and also to promote efficiency. These structures, whilst being good in providing direction and flow of information under stable conditions, could also themselves be barriers in the dynamic environment which is necessary to realise projects. The rigid structures of public sector organisations lack flexibility and adaptability to new circumstances, and, particularly for construction projects whose workload, work volume and resource levels are unpredictable, such structures could inhibit the adoption of partnering.

Another internal barrier to partnering is the cultural attitudes of the organisations forming the partnership. Organisational culture is defined as the pattern of shared basic assumptions that the organisation learned as it solved its problems of external and internal integration, which has worked well enough to be considered valid and therefore to be taught to new members as the correct way to perceive, think and feel in relation to those problems. There is, however, no single correct way for organisations to do business. Thus barriers will be created for organisations that choose to accept only the 'correct' basic assumptions of doing things instead of identifying those that will promote the most successful organisational performance and either maintaining them if they already exist or moving the firm towards adopting them if they do not.

The existing climate of an organisation can also be a barrier to partnering. The climate of an organisation and that of any large project is set by its top executives, and because it is subjective and may often lead to direct manipulation by people with power and influence, executive decisions and policies can affect it. Thus directives from top management that go down the chain of command may significantly increase or reduce barriers to partnering originating in the organisational climate.

Implementation of partnering

Once a contract is agreed, the levels of management who have the authority to enter into the commitment traditionally delegate the managing or 'administration' of the contract to their contract managers (project managers, agents, resident engineers or whatever title they may have). This occurs in both parties' organisations, clients and contractors. Their contract managers are typically two levels down from those who agreed the contract (Wearne, 1992). In this process an agreement to partner needs to be sustained. 'Kick-off' sessions between the parties should therefore be led by the high level of management who entered into the contract and agreed to partnering, so as not only to establish the responsibilities for contract management but also to hand over how their agreement to partner is to operate.

The contract managers should clearly be members of the kick-off sessions and be personally committed to the partnering agreement, but the construction industry worldwide is habitually confrontational and conflict-prone. It has to be accepted that even with a partnering agreement, issues will occur which if not resolved quickly and effectively, could develop into conflict and disputes. Realistic risk allocation in the contract can minimise the occurrence of problems, but if nevertheless a problem arises, one party's contract manager may come under pressure in his/her organisation to exploit it in the traditional way (Bresnen and Marshall, 2000).

The checklist in Box 7.1 should be used in implementation.

Summary

The traditional approach to delivering projects is riddled with conflict and adversarial relationships that result in delays, cost overruns and, in some cases, lower quality.

Construction companies and clients can use partnering to improve their competitiveness, to improve product quality and to keep pace with changing customer requirements. An effective partnering arrangement can replace the potentially adversarial atmosphere of the client–contractor relationship with an atmosphere that will foster a team approach to achieve common goals.

Box 7.1. Implementation checklist

- *Clearly define aims, objectives and long-term goals*
- *Identify a suitable procurement strategy*
- Formality of contract, creation of joint incentives and bonus schemes
- All contract documents are to include a workable framework for the prompt notification and speedy, cost-effective resolution of disputes, which will maintain proper business relationships between the parties
- *Set and carefully plan schedules*
- *Identify potential sources of conflict, disputes and areas that may cause problems* using various management tools, such as risk and SWOT analysis
- *Properly allocate risk*
- *Select and appoint the project team*
- Suitable and compatible staff should be selected. Empowering people is the key to the whole system. This entails adequately training and giving authorities to lower levels to agree matters, but also encouraging open communication and feedback. The team should be driven by shared objectives and goals
- *Clearly define roles and responsibilities and level of authority*
- *Form appropriate problem-solving teams*
- These teams will be responsible for identifying any existing or potential problems, disputes or conflicts, and resolving them, with the cooperation of all concerned in the best interests of the client. Knowledge, experience and innovative thinking should be taken into consideration. Ignoring problems, or no decision, is not acceptable. A problem shall be dealt with as such, without blame or responsibility being apportioned to any one party
- *Introduce 'shadow partners'.* Each member of the team has a member of the opposite team with whom he/she directly relates. These partners have the authority to sort out problems as they arise. The problem will be passed up the managerial line after all attempts to arrive at a prompt solution have failed
- *Partnering workshops* should be held as soon as possible. The purpose of the first workshop is to establish how the partnering firms will work together. It will concentrate on building mutual understanding among all the participants by producing a set of mutual objectives and a problem resolution process. This will then be embodied in the partnering charter

Partnering, however, is not legally binding but depends on trust, commitment and the desire to achieve continuous improvement to thrive. For organisations to be able to apply partnering, a framework must be set up for its implementation. It must also be noted that partnering does not eliminate problems by mere acceptance of its use, but that it demands every player's effort towards achieving results. There are numerous benefits to be derived from adopting partnering but there are also barriers to be surmounted.

Bibliography

Baden Hellard, R. *Project Partnering: Principle and Practice.* Thomas Telford, London, 1995.

Barnes, N. M. L. Civil engineering management in the Industrial Revolution. *Proceedings of the Institution of Civil Engineers,* **138** (2000), 135–144.

Bennett, J. and Jayes, S. *Trusting the Team: The Best Practice Guide to Partnering in Construction.* Centre for Strategic Studies in Construction, University of Reading, Reading, 1995.

Bennett, J. and Jayes, S. *The Seven Pillars of Partnering: a Guide to Second Generation Partnering.* Centre for Strategic Studies in Construction, University of Reading, Reading, 1998.

Besterfield, D. H. *Quality Control,* 5th edition. Prentice-Hall, Upper Saddle River, NJ, 1998.

Bresnen, M. and Marshall, N. Partnering in construction: a critical review of issues, problems and dilemmas. *Construction Management and Economics,* **18** (2000), 229–237.

Construction Industry Board. *Partnering in the Team.* Thomas Telford, London, 1997.

Ellison, S. D. and Millar, D. W. Beyond ADR: working towards synergistic strategic partnerships. *Journal of Management in Engineering,* **11**(6) (1995), 44–53.

European Construction Institute. *Partnering in the Public Sector.* Thomas Telford, London, 1997.

Green, S. D. Partnering: the propaganda of corporatism. In: Ogunlana, S. O. (ed.), *Profitable Partnering in Construction Procurement. Proceedings of CIB (Conseil International du Bâtiment) Symposium,* pp. 3–14. E & FN Spon, London, 1999.

Institution of Civil Engineering Surveyors. Partnering. *Civil Engineering Surveyor,* Oct. 25 (1997).

Kanter, R. M. Collaborative advantage: the art of alliances. *Harvard Business Review,* July–Aug. (1994), 96–108.

Kubal, M. T. *Engineered Quality in Construction.* McGraw-Hill, New York, 1994.

Latham, M. *Constructing the Team: Final Report of the Government/Industry Review of Procurement and Arrangements in the UK Construction Industry.* HMSO, London, 1994.

Wearne, S. H. Contract administration and project risks. *International Journal of Project Management,* **10**(1) (1992), 39–41.

Wearne, S. H. and Wright, D. Organizational risks of joint ventures, consortia and alliance partnerships. *International Journal of Project and Business Risk Management,* **2**(1) (1998), 45–57.

CHAPTER EIGHT

An alliance/partnering contract strategy

G. White

Introduction

This chapter will deal with the development, planning and methodologies used to execute an alliance-type contract strategy (defined as having client/contractor-aligned objectives and using a risk/reward payment concept). However, there is a need to understand and consider why this type of contract strategy was developed for use on offshore oil/gas-type projects. The reasons were primarily to reduce the costs, improve the quality, improve safety and reduce the schedule. The thinking and development of an alliance contract strategy are described in the first part of the chapter. The theory and implementation are then explained through the use of an example of an offshore gas condensate project where such an alliance contract strategy was used. The remainder of the chapter outlines some of the 'key' work processes that were used for the project execution and led to 'best in class' results. It is worth noting that this type of contract strategy can be used in the construction and other industries and in some cases has already been used. Finally, the 'key' work processes can also be used for other types of contract strategies.

Selection of contract strategy

In Chapter 5 a number of contract strategy approaches are outlined, these include alliance, lump sum/turnkey and reimbursable contracts. There can also be a combination of two of these or all three. Some would suggest that what are being referred to here are methods of payment rather than contract strategies. For each contract strategy type, the contractor will be paid for providing services, resources, equipment, facilities, etc. The contractor is then paid for its work, taking risks during the execution and making a profit. The fundamental point is that different contract strategies establish how the contractor makes its profit and how the client minimises the total cost to lead to a successful project outcome. Hence, the contract strategy has a major influence on how the project should be set up, managed and executed. It is therefore recommended that consideration is given to selecting the contract strategy at an early stage in the project planning. The contract strategy includes the following influences on the project:

- project management organisation and team make-up
- project controls to be implemented
- relationships between the client and contractor
- management of interfaces, risks, liabilities, etc.
- achievement of the lowest out-turn cost
- achievement of planned start-up.

The impact of these factors must be fully assessed when setting up a project. For example, the team needs to have the experience and expertise to be able to handle the contractors using the particular contract strategy. Also, the size of the client team, the culture, and the cost and schedule control systems need to be developed to ensure that the costs and schedules provide effective information. Finally, using reimbursable and lump sum contract strategies, costs generally tend to rise after contract award as the client and contractor are on opposite sides. The concept of the alliance contract strategy is to start with a realistic cost target and then the client and contractor work jointly to reduce costs after contract award. The key point is that all the aspects of the project need to be fully aligned with the selected contract strategy.

Commercial goals/objectives alignment

There are three different aspects that have been considered and lead to the use of an alliance contract strategy, as outlined below.

Basic principles

It should be recognised that a contractor will generally make a profit from the contract. Also, it is the client that has a direct influence on the way the contractor makes a profit and on the level of profit that will be made by the contractor, through the selection of the contract strategy. Finally, the contractor's profit should be earned through its performance and not on the contractor's ability to make and win claims.

Key drivers

Usually, and quite rightly, the contractor's method of making profits drives its actions and behaviour and the performance of the contractor's whole team. The client likewise sets up its team to ensure that its concern of minimising the 'total life cycle' is delivered. Finally, it is generally understood and accepted that outstanding disputes are settled by both sides meeting their drivers and objectives, namely, the client ensures that the work is acceptable and the contractor is paid elements of any outstanding claims.

Alignment

The aim should be to try and achieve alignment of the client's and contractor's drivers at the outset. The client needs to articulate what it is trying to achieve in the 'total life cycle' by defining the desired outcome. This could be lowest out-turn cost, early/late revenue and levels of revenue, facility flexibility and/or a combination of some of these drivers and/or performance factors. The contractor rewards could then be based on the expected performance, and the commercial drivers and concept should be developed by working with the contractor community. Work should then move on to defining how profits will be made and risks undertaken by the client and contractors. Finally, the contractual document should be developed to reflect the proposals that will be used to tender for the actual work. It is worth noting that reimbursable and lump sum contract strategies tend not to encourage client and contractor alignment. On the other hand, one of the main features of the alliance strategy is the ability to achieve alignment of the drivers, etc., of the client and contractor.

Alliance principles used on the Britannia development in the UK North Sea

Introduction

Britannia is situated 210 km (130 miles) north-east of Aberdeen, with recoverable reserves estimated to be approximately 3 trillion cubic feet of gas and 145 million barrels of condensate and natural-gas liquids. Britannia's production is approximately 740 million standard cubic feet of gas per day, with initially 70 000 barrels of liquids per day.

Conoco and Chevron formed the first alliance and established a single entity, forming a new operating company, Britannia Operator Limited (BOL). This single entity interacted with suppliers and contractors under authority vested in it by the unit operating agreement. The project adopted the following six key objectives:

- high safety standards
- high environmental and ethical standards

- high reliability and quality
- maximising economically recoverable reserves
- becoming a lowest-cost safe operator
- achieving the lowest commensurate capital cost.

The project was executed using a series of alliances and other types of contract strategies. This section of the chapter focuses on the topsides alliance.

Description of topsides

The topsides are a single integrated production, drilling and living-quarters (140 people) facility with a dry weight of 19 400 tonnes, an operating weight of 28 900 tonnes and 36 well slots. The facilities are designed to handle daily production and to control the subsea satellite. The facilities consist of an integrated deck and a series of modules, and are supported by an eight-legged steel tubular jacket. Engineering commenced in August 1994, with fabrication of the integrated deck starting in August 1995; installation took place in August 1997, with first gas achieved in August 1998.

Development of the alliance contracting strategy

Conventional project execution approaches typically cover the categories of reimbursable and/or lump sum turnkey. Generally, for this size of facility, separate and discrete contracts are progressively placed for engineering/procurement, deck/module fabrication, installation, and hook-up and commissioning. The contract reimbursement arrangements are usually based on some lump sums and all-inclusive rates (i.e. labour costs, overheads, profits, etc.) for the bulk of the work. The bulk of the risk is thus being placed at the contractor's door. The client, with a large team, ensures the quality, handles all the technical and contract interfaces/barriers and carries overall responsibility for success.

For this method of execution, it is critical that the scope of work is well understood, fully defined and fixed, prior to contract placement. The contractors' objectives are thereby aligned to their scope and services, which tend to be misaligned with the other contractors and the client. Also, with this size of platform topsides the individual activities overlap. Hence, design is being finalised during fabrication, which leads to changes and results in cost increases in downstream contracts. The client pays for changes, misaligned interfaces, etc. Finally, these individual contracts can dislocate the total life cycle process.

The alliance project execution strategy was developed as a result of an extensive review of alternative contracting strategies, with a set of principles/best practices being established for the execution of the Britannia project. Also, it was recognised that the platform topsides were on the critical path of the project and all the elements were interdependent, thereby offering an opportunity to reduce costs and improve the management of risks. As part of the development of the alliance contracting strategy, ongoing consultation with industry continued to test the concepts and ideas.

General principles
The project team developed general principles as listed below:

- aligned goals and objectives
- commitment to aggressive capex/opex/drillex reduction
- working to remove process inefficiencies, duplication and traditional contract interfaces
- formation of integrated teams
- early involvement of contractors to optimise design and reduce costs
- aligned and equitable contracts
- shared profits and risks through individual contractor performance and overall alliance performance
- promotion and maintenance of the highest standard of safety and quality.

Concept
Through a spirit of mutual openness and trust, the alliance was formed at the earliest feasible development stage. This was established through ongoing discussions, where the industry and contractors were invited to comment on the completed alliance contract strategy. The basis of the alliance was that by using the above general execution principles, the client and all the contractor members agreed to work jointly to reduce capital, drilling and operating expenditure, meet the schedule, and deliver topsides facilities that met the required quality, operator needs and safety. The financial basis was the implementation of commercial/contractual arrangements, using a risk/reward approach whereby contractor members earn their profits through their performance.

The alliance contract strategy addresses the problems of the conventional execution approach and provides benefits by the following means:

- The client and contractor goals are aligned, whereby cost overruns impact the whole alliance.
- All contractors jointly work on the interfaces/conflicts and share the responsibility for achieving the success of the topsides.

- Contractors are involved early to incorporate their requirements into the design and other associated topsides work to
reduce costs, achieve the required quality, etc.
- Fabricators are involved in the total cost of fabrication and of
hook-up and commissioning (HUC) to reduce or eliminate
carry-over of offshore work.
- First-year operations performance failures are offset against the
alliance profit.

The contractors were required to tender the expected out-turn
total costs and to accept earning profits through cost reductions and
the need to achieve the client's facility performance requirements.
This produced a culture enabling costs to be reduced (e.g. rather
than using changes to gain profit) and total contractor buy-in,
coupled with the contractors' ability to influence the topside development on the basis of the client's requirements.

The problems with this approach are:

- selecting and committing to contractors at an early stage of the
project
- managing changes
- changing the traditional industry culture, attitudes and work
practices
- ensuring that the quality, total life cycle and operator requirements are fully understood, implemented and met
- retaining effective and timely decision-making capability.

These problems were resolved within the Topsides Alliance execution using the work processes outlined later in this chapter.

Formation

The Topsides Alliance was established in late 1994 and was made up
of seven members, with each member having a specific role and
responsibility. The alliance was led by the BOL and AMEC (Design)
alliance members and was headed by a division manager. He was
supported by design, construction, project services and procurement managers, who together formed the Topsides Management
Team (TMT). This group was accountable for the day-to-day
management of the topsides and other support services. In addition,
there was the Topsides Alliance Management Team (TAMT), which
was made up of the above managers, the senior full-time representatives of each of the other five alliance members and a representative
from the client's operations group. This latter individual acted as
the ultimate customer, contributed to all aspects of the development and accepted the completed topsides facilities on an ongoing
basis. The TAMT was accountable for:

- overall direction
- meeting the topsides objectives
- ensuring the execution philosophies were implemented
- ensuring quality requirements and safety targets were met.

Finally, the Alliance Board comprised senior representatives of each alliance member company and was chaired by the client. The Alliance Board monitored the progress, costs, schedules and major changes of the Topsides Alliance.

Commercial arrangements

The commercial arrangements for each alliance member were similar and were typically broken down into four elements:

1. *Lump sums* – for 'non-scope-dependent' work covering items such as yard facilities, corporate overheads and computer-aided design (CAD) facilities.
2. *Reimbursable direct costs* – for 'scope-dependent' work relating to management, supervision and labour. The contractor was reimbursed for all resources used and work carried out. This cost was defined as a 'base cost' for establishing the performance incentive. Also, it was assumed that the 'base cost' represented the out-turn cost of this type of work. Finally, all fabricator-supplied materials and external services were paid at cost.
3. *Performance incentive* – this was the profit or loss for the individual contractor's own effort in reducing the 'base cost'. The performance incentive was set by each contractor company, at the tender stage, and reflected its view on risk and on reward for its own performance. This element of profit represented between a quarter and a half of the potential profit that an alliance member was likely to earn.
4. *Alliance incentive* – profit based on performance of the whole Topsides Alliance against the Topsides Alliance capital costs. The alliance incentive, which represented the balance of the total profit, was established by deducting from the Topsides Alliance capital cost target the total out-turn cost invoiced by all the alliance members including the performance incentive, the out-turn costs of all the equipment and material, and other work and services. The alliance incentive was then distributed against the agreed fixed percentage shares.

Contract

The entire contractual agreement, which was between BOL and each of the contractors, was based on a Conoco conventional contract. The contract had been enhanced to include the alliance arrangements, and addressed some of the problems identified, as follows:

- The standard-of-workmanship guarantee was deleted, except for wilful misconduct. Any rework was paid for and added to the total costs, thereby reducing profits.
- Claims for extras due to changes in work could only be made if changes were made to a limited number of principal design parameters of the facilities. (That is, the basis of design was not part of the contract.)
- Alliance profits were paid one year after production start-up, and only if the acceptance criteria had been achieved. The acceptance criteria, set by the client, covered meeting operating costs, production and productiveness targets, gas customer performance requirements and the first-gas date.

Alliance charter
The alliance charter was a non-legal document of intent, developed to further amplify the alliance execution principles and to align and encourage joint working and promotion of relationships. This charter was signed by alliance members' very senior executive officers, thereby stating their commitment to the success of the Topsides Alliance.

Alliance execution
The overall accountability for the management, direction, budgets, control, etc. of the topsides was with the TAMT, specific elements having been further devolved to individual managers and senior team members. Reporting on costs, progress, etc., for the topsides work was performed by a central group, which received its data from individual groups; the reports were issued to the alliance.

Solving the initial problems and turning the alliance into reality required a major investment of time, and needed different methods of working, new processes and team-building sessions, with the recognition and belief that profits can be achieved through cost reduction. The work processes developed are outlined below.

Work processes to make an alliance contract strategy happen

Individual and team culture
The alignment and culture of individuals and teams had a number of different stages as follows.

Individual alignment and enrolment

The first stage was to inform people and explain the project, covering the concept, organisation, alliance commercial arrangements, etc. This was followed by outlining the required cultural changes that the individual was expected to make, and included the need to manage through influence rather than, say, control, remembering that the contractors were part of the risk/reward scheme and therefore had a voice, and always promoting and working to the alliance principles. This led to the individual having to make a decision – is he/she willing to make the required changes?

Team-working and enrolment

This was a critical process, which was designed to provide leadership for alliance behaviours and work process improvement throughout the topsides work, by giving topsides team members the appropriate skills and competencies. This covered the concept of orientating the whole team with the alliance culture and was executed in several stages.

- *Stage 1* – a two-hour session on the first day covering the project, alliance, commercial model and project safety.
- *Stage 2* – after about six weeks, a one-day session building on initial experience and extending to fully cover all aspects of the alliance and processes, addressing the problems and issues.
- *Stage 3* – this was followed much later by further training of the leaders and individuals who could influence a group of people but were not necessarily managers or supervisors. This involved a 1–2 day session with a training company and covered the following:
    - Development of team skills and behaviours of topsides 'influencers', who were senior project team personnel. Through the use of skilled trainers, the intent was to make these individuals more effective line facilitators, thereby endeavouring to provide leadership through facilitative skills.
    - Provision of support to line facilitators through management sponsors.
    - Provision of chargehands/foremen at the fabrication yards with the skills and behaviours to foster open communication between teams.
    - Provision of additional external facilitation support for one-off key events and/or major problem-solving sessions, etc.

This plan made a major contribution to the achievement of the cost reductions, to meeting and endeavouring to exceed the safety targets, to promoting a 'step change' improvement in all activities and to ensuring the 'seamless' progression of the project through each of the key phases. It also enabled the individuals to make use of these additional skills and competencies on later projects.

Project and work processes

Many project and work processes were developed to enable the objectives and commercial drivers to be achieved. Some of the processes are described below. These processes can be adopted, changed and applied to many other projects.

Physical integration of the topsides team

This involved taking fabrication and HUC team members and physically locating them within the engineering/project services team to be part of the design development, and provided guidance to the designers to achieve the lowest out-turn cost. The intention was to work as a joint team, sharing objectives and commercial drivers, minimising design rework and approval cycles, etc. This arrangement is shown diagrammatically in Figure 8.1.

Alignment of design/procurement and fabrication

It was recognised that every fabricator had its own methods, processes and ideas for fabrication, and it was important to get their ideas for cost reduction. A general process, shown in Figure 8.2, was developed to plan the work and draw out the contribution from the fabrication team.

This process also brought in the HUC alliance member and the installation contractor, who was not part of the alliance, but whose input was vital, since there was a need to lift a deck weighing over 10 000 tonnes, and to lift very large modules to a great height. However, one of the most difficult lifts was a 95 m long, 700 tonne flare boom, which required both cranes of the heavy-lift vessel to be utilised.

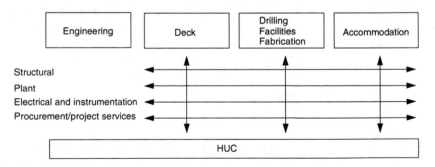

Figure 8.1. HUC team members

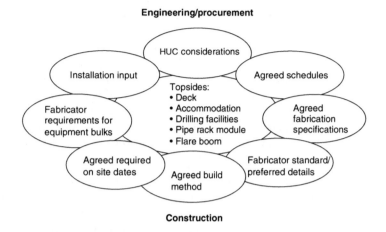

Figure 8.2. Process diagram

Engineering and procurement deliverables

Each of the fabricators specified in detail all the different types of deliverables that were required. The joint teams then investigated the most efficient way of producing the deliverables. This led in some instances to the designer undertaking additional work to produce more detailed types for deliverables that included, for example, shop drawings and detailed piping and instrumentation diagrams. Also, the fabricator was involved in all aspects of procurement to help with sequencing, help to reduce waste, help to allow more engineering time due to reduction in required float, etc. Finally, the IT group set up a system to enable the fabricators to have access to the 3D design model being developed by the design team so as to work more closely with the designers, follow progress, help with clashes, etc.

Level 2½ integrated schedules

Generally on a major project there is a hierarchy of schedules, as follows:

- level 1 – overall project master schedule
- level 2 – major element of the project, e.g. platform topsides, jacket or pipeline
- level 3 – platform topsides engineering, procurement, module fabrication, etc.

It was felt that within this type of hierarchy it was not possible to optimise the detail engineering, procurement, fabrication, etc. of the deck, modules, accommodation, etc. To solve this problem, level 2½ schedules were developed; these are shown diagrammatically in Figure 8.3.

Each of the schedules for the deck, drilling facilities, accommodation, etc. was developed and agreed jointly by the relevant alliance individuals as a team. Then individual engineering/fabrication managers accepted joint accountability for making it happen. These schedules were used to track the overall and critical activities of the deck, etc. From this level, the team were able to drill down to levels 3, 4, etc. to identify the root causes of problems and rectify them.

Topsides mechanical completion and commissioning

The same approach was taken as for the level 2½ schedules above; this is shown diagrammatically in Figure 8.4. The schedules were developed using a backward pass starting from first gas, with the intention of making the project commissioning-driven in the latter stages. Also, the goal was to achieve the lowest-cost solution for the topsides and Britannia between the onshore and offshore work. Finally, the commissioning was integrated into hook-up, mechanical completion, etc.

Occupational safety

This aspect was considered to be a key driver to improving construction productivity and saving of costs, both onshore and offshore. If the total costs of an incident were taken into account then it was soon realised that accidents were very expensive and time-consuming. For example, if an incident took place, consider the costs associated with a 12-hour stop in production to investigate the accident. The major activities and initiatives that were undertaken to improve occupational safety are outlined below.

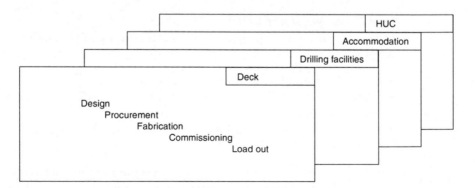

Figure 8.3. Level 2½ schedules

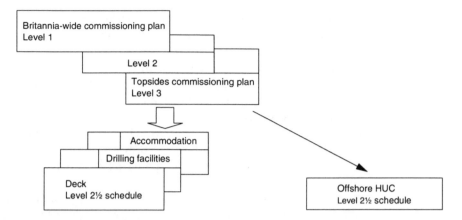

Figure 8.4. Topsides schedules

- *Topsides construction safety group.* Chaired by the topsides construction manger, the group were required to:
  - develop standards
  - share and develop solutions
  - review and recommend improvements to existing practices
  - initiate new ideas.
- *Management supervision and individual commitment through leadership, action and financial reward.* 'Walk the talk'.
- *Continuous communication of the importance of safety, adjustment of the message to suit circumstances and fabrication stage, and use of initiatives.* This meant:
  - one-day training of the total workforce
  - two-day training of management and supervision
  - safety initiative scheme based on unsafe acts, housekeeping, etc.
  - specialist training to include accident investigation, risk assessment, etc.
  - continuous learning from accidents, near misses, etc.

The effort meant taking time out during the day, every day, putting safety on the 'to do' list, etc. to achieve a real step change.

Producing the results

The major features/drivers needed to achieve a successful outcome include the following:

- Assemble the operator team with all the contractors as early as possible, preferably before project sanction and during preliminary engineering/front-end engineering and design.

- Ensure that the operations group are also brought in and treated as the client/customer, being defined as the group who will operate the platform after first gas/oil.
- Develop the alliance principles, commercial model, management principles, etc. in dialogue with the industry to ensure that the industry will be interested in participating out of choice rather than necessity.
- Using the project requirements, benchmark the commercial and schedule targets as the starting point for setting up the alliance and ensure that there are challenging targets for the alliance to achieve.
- Once the alliance is set up, begin to develop very challenging targets using fabrication and HUC alliance members by:
  - aligning the design
  - optimising the overall schedule for the design, procurement, fabrication and commissioning
  - ensuring that the design suits the fabricators, HUC team and operations team.
- Fundamentally, designers and engineers are required to produce a design for the service condition meeting all the safety factors, etc., and to ensure that it can be built. The latter need should be very important and can make a major contribution to reducing costs without compromising quality.

When the project is under way and problems begin to occur, it is absolutely vital that the alliance use the established and agreed principles, processes, etc. to solve the problems.

Note: if this point is violated the whole alliance will begin to fail and will never recover. *Staying focused and retaining the high ground will deliver the targeted result.*

Quality and safety

The general philosophy is to adopt management systems based on goal setting, continuous improvement and the use of joint teams with all the skill sets and competencies to provide assurance of actual delivery.

Thus, the next step is the setting up of safety and quality management systems. Also, the engineering and operations group should be used to contribute to the targets and goals. The important point is that, for example, the level of inspection needs to suit the design. This should ensure an optimum and appropriate inspection level and may mean a greater level of inspection for particular equipment, materials, etc.

Finally, a culture of continuous improvement should be encouraged, as exceptional safety and quality performance does lead to improved productivity, beating the schedule and lowering out-turn costs.

Summary

It important to begin to think in terms of what is possible in the future based on what has already been achieved to date.

The concept of continued commitment to safety should be part of all future projects. This is because safety should be considered of equal importance to production, quality and cost reduction/profits. A good safety performance is as important as any of the other key performance indicators for a successful project.

From a client perspective, it is important to take on board the concept of achieving the lowest total life cycle cost, which includes prospect-finding, capital and operating costs. The industry can make a significant contribution to these goals and there is a potential for many and varied contributions. The alliance strategy can underpin the achievement of the lowest cost though the adoption of appropriate equipment quality and improvement of operability and maintainability, though developing the appropriate alliance commercial models.

New ideas need to be encouraged to reduce costs and provide value-adding solutions. By the involvement of contractors early in the life cycle, with and/or through an alliance arrangement, and by solving problems in the early development phases, a marginal prospect could be turned into a project.

Part of the way forward is for everyone to improve their ability to deliver on their commitments. The way forward also involves working together to become more interdependent, and to be rewarded through performance in achieving the aligned, agreed goals rather than through the use of claims, delays, etc. through inequitable contract arrangements.

To date, the alliance strategy has only had some success. The failures were partly due to not following through a systematic approach and the subsequent execution, etc. However, by turning this strategy into a dynamic process and through continued improvements the present achievements can easily be improved beyond normal expectations.

This type of contract strategy is transferable to the construction and other industries where clients have to engage contractors to execute projects, and similar or better results can be achieved.

CHAPTER NINE

Joint ventures

D. Wright

Introduction

This chapter discusses some of the possible implications for the client of dealing with not a single contractor, but a group of contractors acting as a single entity – a *joint venture*. This entity may take the form of a jointly owned company. More probably, it will take the form of a 'partnership' within the true meaning of the law, a group of people or companies who have agreed to carry out a joint commercial activity for profit. Dealing with such a partnership undoubtedly gives the client several advantages, but it also has certain drawbacks. (The joint venture will have more skills and resources than any one of its members. That gives it more capability – but also more bargaining power as well.)

Some drawbacks will be obvious. A joint venture will probably be rather more ponderous during contract negotiation. It may take longer to submit a tender. However once the contract is agreed, provided the joint venture is properly organised, there should be a central project management team in place for the client to deal with so that the contract can be run normally. Nevertheless, there are risks present when dealing with a joint venture that are not there when dealing with a single company. The joint venture is more fragile than a single contractor. It can be disrupted by the loss of one of the partners or by dissension between them. This chapter therefore looks at some of the organisational problems that can occur within a joint venture and how they might be avoided.

Background

Sooner or later every large organisation will find itself buying from a 'joint venture' – a combination of commercial organisations, companies, partnerships and others, collaborating to take on a significant contract (or project). The usual reason why the contract will be significant is that it will make demands that a single contractor will find difficult to meet from within its own resources or through normal procurement/subcontracting. Typically this will be because the contract is too large for the contractor to manage comfortably alone, or too complex, or both.

Indeed, the trend is for large projects to become more common and for large projects to become larger. Even if they are not becoming larger then they are certainly becoming more complex, involving a wider range of engineering and management skills and disciplines, and taking longer to carry out.

Over recent years we have even seen the emergence of the ultimate long-term complex project, the 'construct and then operate' project (build–own–operate, build–operate–transfer, build–own–operate–transfer, etc.), as epitomised in the wide range of private finance initiative (PFI) projects coming into being within the UK and the public–private partnership contracts in the UK and other countries. Such projects, and the contracts that they require, are certainly long-term, requiring time to negotiate and to build or construct, perhaps then followed by an operating period that may run for several years. They are complex both because they often require a wide range of engineering skills and disciplines during the build/construction phase, and because they may then require totally different skills during the operating phase.

The long-term project and the complex multi-disciplinary project present both client and contractor with special problems.

For the client, the main problems are perhaps those of initial project definition and then the inevitable changes in need/requirements that will arise. Complex projects or systems are easy to define in broad terms but not easy to define in contractual terms. The contract demands relatively precise task definition. Precise task definition then ensures precise provision by the contractor – at least that is the theory. Unfortunately, in real life precise task definition may not really define what the client actually needs and, in the long term, the client's requirements can never remain constant anyway.

The long-term, multi-disciplinary project also creates special problems for the contractor in providing the skills necessary to carry out the different tasks required by the contract. A wide range of skills may be required during the 'construction' phase. Where that construction phase is then followed by an operation phase, the problem is even worse. Operation often requires totally different skills and people from those required by construction.

Clearly, as a project becomes more complex it becomes harder for any single organisation to contain within itself all the skills and resources that are necessary to carry out that project. Therefore the contractor will always need to find skills and resources from outside if it is to complete the project.

One obvious way of providing those skills and resources is by subcontracting, placing contracts with appropriate subcontractors to buy in from them work and skills resources or equipment that the contractor itself may not be willing or able to provide. Every contractor will use subcontracts. They are useful. They can and do provide resources or equipment. But there are limits to what they can contribute to the project. Every contractor will understand the problem of creating 'project commitment', or 'project risk/benefit sharing' in its subcontracts. Every contractor will also know the problems created by the 'major subcontract', where the contractor is compelled by circumstances to place one or a small handful of subcontracts that are so large in relation to the remainder of the project that they can come to dominate the entire project. Finally, every contractor knows the problem of the 'critical subcontract', the subcontract to buy in a single item of equipment or a skill/resource which the contractor cannot provide for itself, but which is absolutely vital to the overall success or profitability of the entire contract.

The common factor in all these for the contractor is that of trying to create commitment to the success of the main contract/project. Subcontracts create subcontractors. Subcontractors are committed to the success of their subcontracts, rather than to the success of the main contract. Obviously they will do their best, and many subcontractors' best will be very good indeed. But in the final analysis their main concern has to be the profitability of their individual subcontracts for themselves, not the success of the project or the profitability of the main contract for the contractor.

These are the sorts of considerations that lead contractors into joint ventures, collaborating with other organisations more or less as equals, acting as joint 'partners' in one or more projects, and sharing the overall profit/loss and risk between them.

Of course, there are a number of other possible advantages to contractors in the joint venture, quite apart from the pooling of resources and project commitment.

A joint venture allows the partners to share the preliminary work and costs of a project, such as tendering, negotiation, research or development. It also allows the partners to pool commercial contacts or knowledge of the client or market. Sometimes they may even share technical skills or know-how.

Clients often want to deal with a single organisation. On a major project this can sometimes only be achieved through a joint venture. Against this criterion, a joint venture has a more powerful base in negotiations with the client, as well as with government organisations, banks, suppliers and others.

For the client, dealing with a joint venture is a more high-risk strategy than dealing with a single contractor, in the sense that the joint venture always adds an additional dependency risk into the equation, that of dissent between the partners, or even the total breakdown of the joint venture.

Therefore the client has an additional consideration to bear in mind. The client needs a good contract. It needs competent contractors. It also needs to deal with a 'good' joint venture – in the sense of a joint venture with survival/staying power, comprising companies/organisations that will cooperate effectively with each other and work efficiently together for the lifetime of the project. Every client should question a joint venture, to ensure that it is properly put together. Of course, any joint venture can suffer internal conflict; but the better the organisation of the joint venture from the outset, the lower the risk.

Types of joint venture

A *horizontal* joint venture is one in which the partners can carry out their work, more or less, in parallel and at the same time. This creates a straightforward relationship in which all partners depend upon each other and all partners will contribute to and profit from the project at the same time as each other.

A *vertical* joint venture is one in which the activities of one or more of the parties are not interdependent, but follow on from each other in sequence. An example would be the building of a toll bridge by some of the members of a joint venture, which was then to be operated by another partner. Here the work of some members of the joint venture would be complete before the work of the operating/management partner could even commence. Relationships in a vertical joint venture are obviously rather more complex (and perhaps rather more subject to stress).

A *homogeneous* joint venture will consist of partners drawn from the same industry, or industries that are related to each other. For instance, a joint venture made up of companies within the building and civil engineering industries would be homogeneous. So would a joint venture made up of chemical engineering and process companies. A homogeneous joint venture is again straightforward, because the partners operate within the same contract and technical disciplines as each other. They normally undertake similar classes of work and risks, and therefore understand each other.

A *heterogeneous* joint venture is made up of partners from different industries or disciplines, who therefore normally undertake different classes of work. This type is rather more complex to create and perhaps rather more difficult to manage on a day-to-day basis than a homogeneous joint venture, but with the right attention to reporting/management/communication aspects, will operate perfectly satisfactorily.

(In addition, of course, any joint venture may be *national*, involving partners based within one country, or *international* involving partners based in two or more countries. Obviously enough, any international element in a joint venture simply increases the culture/communications problems within the joint venture – but of course may add considerably to the abilities of the joint venture.)

Any individual joint venture may be both horizontal/vertical and homogeneous/heterogeneous. In theory the most difficult to set up and manage is the vertical/heterogeneous joint venture, especially when it is international in composition. In practice there may be very little between them provided that the partners have put the joint venture together properly and then commit adequate resources to operating within it.

Formation of a joint venture

Basic rules

- *Festinate lente*, a Latin proverb – 'Hurry slowly'.

When a joint venture is put together there is a lot to be agreed. Quite apart from this, the partners also have to make sure that they know each other, understand each other and trust each other. This takes time. Also, some aspects of the deal cannot be hurried. For instance, it is a matter of moments for the partners to agree a principle, say, that they will adopt the same costing/pricing basis, to ensure that all parties make approximately the same level of profit on the basic work and on any extras and variations. However, working out the practical implications involved and comparing the actual costing methods used by each of the organisations involved, even within a homogeneous joint venture, will demand probably several meetings spread over weeks or months.

- 'If they can't all share the same taxi there are too many of them.' (Anon.)

Every large project will involve work and equipment supplied by many different organisations. However, they cannot all be members of the joint venture. Some joint ventures operate completely successfully with a comparatively large number of partners. However, the basic rule for the vast majority of joint ventures has to be that the number of partners should be limited to a maximum of four or five.

The reason for this is simple. To assemble the joint venture requires detailed discussion between the partners of a whole range of difficult issues concerning obligations, risk and money – usually against a deadline. Anyone who has ever been involved in the process knows just how difficult this can be, even when only two or three companies are involved. If there is a large number of partners that has to be involved in the negotiations then the chances are that those negotiations will be fudged. If the same large number of partners is then involved in carrying out the project and there are problems then those problems will not be easy to settle.

- Partners in a joint venture should have an approximately equal (financial) stake in the contract.

It can often appear to be very advantageous to include as a partner in the joint venture a specialist supplier, such as a design partnership, because that supplier's work is seen as important to the success of the overall project. Indeed, we have already commented that one of the difficulties of subcontracting is that it does not deal properly with that sort of subcontractor. However, the basic rule should be as stated above. The reason is that the joint venture partners have to accept risk as equals, which a small participant usually finds difficult, if not impossible, to do.

Establishment
The following principles are largely self-evident, but worth repeating:

- Every partner should begin by reaching internal consensus, both as to its own objectives for the joint venture as a whole, and for itself within the joint venture.
- Before any detailed negotiations take place there must be a general discussion, at senior management level, to ascertain whether there are differences between the objectives of the partners. If there are differences and these cannot be reconciled then the negotiations should not proceed.
- All negotiations must be conducted at the appropriate level for achieving commitment.
- Naturally, the negotiations must deal openly and honestly with the practicalities of the contract/project – the work to be done, the difficulties and the attendant risks. From this should follow agreement on the share of work and risks between the partners, the interfaces between them, a realistic programme/work schedule, a realistic policy for dealing with the problems that will inevitably arise, and the respective responsibilities and liabilities of the partners.

- The negotiations must seek to identify and then eliminate potential causes of deadlock between the partners wherever they may be. Negotiators must then concentrate upon identifying differences between the partners that might cause problems later on, so that they can also be eliminated in advance. The aim must always be to minimise any potential for conflict between the partners.
- The negotiators must accept, however, that differences will still arise and build in procedures to deal with those differences. Differences are inevitable, but they must not be allowed to become disputes. Differences must also be able to be settled quickly – a long-running difference is a dispute.
- The negotiations must put in place a 'management structure' for the project. They must also put in place a management structure for the joint venture. (The relationship needs managing just as much as the project.)
- The greater the time that can be allowed for negotiation the better. The partners need time to get to understand each others' problems and attitudes if they are to collaborate successfully.
- Trust may need to be established between the partners' representatives in the negotiations. (This itself may require time to achieve, probably quite some considerable time.)
- In addition there may be communication difficulties to be overcome, caused by differences in language, culture, geography, disciplines or experience.
- The partners' representatives should include at least some of the personnel who will be involved in managing the project.
- The aim is to create first understanding, and then collaboration.
- Negotiators must avoid a 'win or lose' attitude. Almost invariably, in a joint venture, if any one partner 'wins', all lose in the end.
- The results must define every partner's aims and commitments.

The common organisational risks

The aim of a joint venture is understanding, teamwork and collaboration. The target when setting up the joint venture is to foresee and agree what relationships and commitments between the partners are needed to control and carry out their external commitments to others, together with the internal mechanisms to deal with differences between the partners as they arise. The two things that can be absolutely guaranteed within any joint venture are that there will be differences between the partners from time to time, and that some of those differences will be totally unforeseeable until they arise.

A joint venture generally involves more causes of risk than other types of project or business organisation. The extra causes depend on whether the partners really are willing and able to collaborate with each other in sharing the problems and risks involved, and whether they have made realistic arrangements that will enable them to do so in practice.

What follows is a list of some of the problems that tend to occur on a regular basis – with suggestions as to how to avoid them. Naturally, most problems can be managed or tolerated. However, experience shows that anticipation and avoidance are usually better than toleration in the end.

Differences in objectives

The partners may differ in their understanding or interpretation of the objectives of the joint venture, and this may not be apparent before the joint venture has entered into commitments to the client.

The partners' reasons for entering into the joint venture, together with their objectives, should be clear to all before the joint venture is agreed.

Changes in objectives

A partner's objectives may change during the project due to external factors, such as a merger or takeover.

All partners must accept that the joint venture agreement is a binding commitment for the duration of the project. (The only permitted way out is, usually, that any partner will be allowed to quit the joint venture if a change in circumstances makes it impossible for that partner to continue. However, the partner must find an acceptable substitute to carry out its remaining obligations, and compensate the other partners for the costs of the disruption that it has caused.)

Communication problems

In theory any group of contractors can form a joint venture that can be successful – however varied the partners are in size and skills. In practice different industries, and the companies in them, tend to live in different worlds and therefore have difficulty in achieving understanding and easy communication between them.

Management and communications procedures and systems need to be thought through from the beginning. In addition, special care needs to be taken to ensure that the management personnel from the different partners meet regularly, probably both on a formal basis and informally. The cost of this must be allowed for in the project costing.

Changes in demand

The needs and risks will change during the project, often because of matters outside the partners' control. The most common reason will be changes in the requirements of the client.

Change within the project can affect the balance between the partners. It can make the project more profitable for one partner at the expense of another, for instance. This has to be recognised from the outset and procedures put in place to deal with the problem. A simple agreement to collaborate may be sufficient to start a joint venture, but will never cope if change disrupts the relationships between the partners. When forming a joint venture, the partners should agree how its structure is expected to evolve as relationships become more complex during the project.

Project termination

Sooner or later every project comes to an end, through termination or when the work has been completed. When that happens the joint venture will also probably need to end.

The joint venture should not commit itself to a project or a contract until the partners have agreed what their responsibilities are to be, the system for managing the joint venture's commitments through to completion, and the procedure for dissolving the joint venture when its work is finished and the remaining assets and liabilities need to be shared out between the partners. Joint ventures are easy to start, but hard to end happily.

Divergence of interests

It is easier to start a cooperative venture than to sustain it. The risk of gradual divergence of interests between the partners is always there, and increases geometrically as the number of partners rises.

A separate management team authorised by and reporting to the partners as a whole, possibly through a management committee, may be needed. It can provide central contract/project management on behalf of the joint venture towards the client and also hold all the partners to their commitments and responsibilities. Of course, this management team must be able to instruct the partners as to what is required to meet their commitments to the joint venture. The partners have to accept that they are subordinate to their own creation.

Provisions for risks

Partners may fail to make adequate cost/budget provision for the greater expense and risks involved in operating within a joint venture. Inexperienced partners, in particular, often greatly underestimate (or sometimes even greatly overestimate) them.

Costs and risks should be assessed by the partners jointly and then allocated between the appropriate activity centres, to avoid double budgeting.

Balance of work

In the normal way of things, the joint venture will be only one part of the business activities of the partners. For one partner it may be a major part, but for another it may only be a small part. Changes to the order books of the partners can affect the view they take of the joint venture.

It is necessary to anticipate conflicts of interest in planning and controlling the work of the partners, and this is usually done through the management team. (The problem is usually easier to control if the joint venture is homogeneous and/or horizontal. It is also easier if the partners share other interests.)

Subcontracting

Uncoordinated subcontracting can cause various problems and conflicts.

Of course the partners in a homogeneous joint venture are likely to understand each others' work and cooperate in overcoming problems rather more easily, but subcontracting may need prior agreement on policy and central control.

Cultural attitudes

The partners (especially in a heterogeneous or international joint venture) are dependent upon each other, but may also have insufficient understanding of each others' work and internal culture.

There is always a need for a formal system of planning and control within the central joint venture management team as well as within each of the partners.

Project attitudes

The partners can vary in their experience of joint venture projects and risks, resulting in differences in the real authority and attitudes of their representatives.

Every partner in any joint venture should be represented on the steering group by a director of the parent company. (This is especially important if the joint venture is heterogeneous or international.) The only thing that can overcome problems such as these is regular contact at senior management level.

Contracting authority

The joint venture must be able to act with authority, and separately from its partners.

One way of achieving this is to entrust the management of the joint venture management team to a manager, essentially the overall joint venture project manager, with authority to act on behalf of the joint venture as a whole. This person may then report to the joint venture steering committee. This has several functions. These would usually include acting as a court of appeal to resolve differences between partners, and acting as the management board for the joint venture. The steering committee would meet on a regular basis, and would comprise senior management representatives of each of the partners. Two-tier management committees have also been used successfully in heterogeneous or large joint ventures, the upper tier consisting of senior management who are authorised to take risks and settle disputes, and the lower tier consisting of the operating managers who control the partners' resources required for the joint venture's work.

Control in default of planning

The need for central control of a joint venture project or contract may become accepted only when policies are not proceeding as intended.

Care must be taken not only to institute the monitoring and managerial decisions needed to carry out the joint venture's commitments, but also to correct any more fundamental failure to plan ahead and anticipate problems. Control can be exercised in various ways, depending upon the structure and relative interest of the partners.

Collective management of problems

The joint venture may need management styles and systems different from those used by partners in their normal business. As in any committee, the partners' representatives on the steering group may run the risks of discontinuity in their knowledge and attitudes about the joint venture business, and tend to 'group think' or be over-cooperative in relation to their parent enterprises' interests and commitments to the joint venture.

The potential risks of joint decisions may be reduced considerably by selecting senior managers/representatives with organising and negotiating skills. The more complex the relationships in a joint venture, the greater may be the need for more analytical and systematic use of quantitative data to analyse problems and choices.

Management quality and motivation

The joint venture needs to be equipped with managers at least comparable in ability to their opposite numbers within the partners. There can be conflicts between joint venture and partner project managers, not least because of their different roles, objectives and accountability.

If the joint venture is only temporary, as is common, its managers need authority from the partners' steering group. The steering group may also have a role to play in supporting the joint venture project manager against their own project managers. Managers seconded to run a joint venture from the partners should have a future in their parent organisations that will be enhanced by success in their performance for the joint venture.

Risk awareness

Few individuals work in more than one joint venture in their career and therefore can bring experience into another, so many managers will be new to a joint venture and therefore not aware of the risks.

The prospective partners to a joint venture should adopt a deliberate policy of searching out available advice and experience to identify and assess the potential risks and possible remedies before commitment.

Incorporation or cooperation?

As was noted at the beginning of this chapter, a joint venture may take the form of either a simple partnership, a group of organisations working together under the terms of a contract between them, or a separate company jointly owned by those organisations. The company operates as a legally separate entity independent of its owner partners, but there should still be a formal agreement between the partners as to the form that the subsidiary company will take, its scope of activity, and the way in which it will be controlled and then brought to an end.

The potential additional benefits to the partners, and perhaps also to the client, of a incorporated joint venture are:

- It can provide a limit to the liability of the partners. (Therefore the client might well require parent company guarantees, for example, from the partners.)
- It may provide longer-term (though not short-term) flexibility in capital structure.
- It may be able to obtain trading benefits that might not be open to each partner alone.
- Because it has a separate organisation it may also be able to provide greater continuity of staffing.
- It may have a better presence in the marketplace, particularly in export markets.
- It may have greater credibility when dealing with government, financial or other organisations.

The attendant disadvantages are:

- It may require more time, effort and expenditure to set up.
- It may be less flexible in the distribution of profits and losses.
- It may result in double taxation.
- It may have less power when dealing with its owners.

Incorporation is therefore preferable for a project or series of projects which require the long-term application of management and other resources. A collaborative agreement is preferable when carrying out a single project or contract. It is more flexible in management, capital structure, and profit and loss distribution. Because such a joint venture is not a separate entity the client can be rather more in direct contact with those with doing the work for its project.

Complexity and control

The client, and any financing body involved, will normally require the performance of an incorporated joint venture to be guaranteed by the partners. If the joint venture is simply agreement-based then the client will either require the contract to be signed by all the partners or require that the non-signing partners will guarantee their performance.

When the joint venture is incorporated, some work may be carried out by the joint venture itself. All other work will be subcontracted by the joint venture to the partners. When the joint venture is agreement-based, either the partners will distribute the work between themselves by a series of subcontracts, formal or informal, or one partner will act as 'lead contractor' and place subcontracts with the rest.

While the client should always insist upon a single channel of communications with the joint venture, usually through the joint venture's project manager and management, control of the partners' operations need not be centralised. Horizontal, homogeneous joint ventures usually need the least centralisation of management, unless international considerations affect the issue. Vertical or more complicated joint ventures may require formal linking and overlapping/duplication of the partners' systems of organisation. This need not be on a large scale, but must be there from the start.

Formal delegation of authority to a separate dedicated joint venture project manager is always important. It is almost essential where the partners are of diverse types, say a combination of commercial companies and public authorities. It then avoids the possibility of differences in corporate autonomy and accountability causing operational differences.

Summary

The terms and organisation of any joint venture will be decided by the partners and will depend upon their relative strengths and interests. No two joint ventures will ever be the same.

The client will never know for certain what the terms are or whether the organisation will stand up to the demands of the contract or the pressures of the project. Nevertheless, the client must try. Good management practice is the same when applied to a joint venture as with everything else. It is prudent to anticipate potential problems and to agree clear arrangements for resolving them before they happen. Everyone involved can then organise accordingly. A good client will, as they say, endeavour to check this out before contract. A good joint venture will be happy with this.

The greatest single lesson of past experience with joint ventures is that the partners should not enter into the joint venture until they have already agreed how it is to be organised and to operate in carrying through its activities to completion.

In broad terms, the structure of the project should maximise collaboration but minimise operational interdependence between the partners. This might not be what one or more of the partners would prefer, but is advisable. This should be anticipated when the joint venture is created.

The style and system of management appropriate for cooperative concentration on the joint venture project are likely to be different from the styles and systems used in the partners' organisations for their normal business.

Most of the risks and problems are predictable and can be avoided or controlled by reasonable forethought. Companies looking to engage in a joint venture should therefore adopt a systematic and logical approach to take into account the nature and organisation of the joint venture itself. Clients dealing with those joint ventures should then check that they have done so. The joint venture can be an effective system between enterprises. It is potentially strong in pooling resources and expertise but weak in its possible divergence of interests. The joint venture is therefore always potentially unstable unless properly assembled at the start and properly managed from the start.

Bibliography

Andreae, N. Joint ventures in France. *Joint Ventures*, May (1992), 31–37.

Benitez Codas, M. M. Cultural integration in bi-national joint ventures. *Project Management without Boundaries, 11th Internet World Congress on Project Management*, pp. 155–164. Florence, 1992.

Boeva, B. Management of joint international projects. *International Journal of Project Management*, **8**(2) (1990), 105–108.

Ellison, J. and Kling, E. *Joint Ventures in Europe.* Butterworth, London, 1991.

Jaafari, A. Management know-how for project feasibility studies. *International Journal of Project Management*, **8**(3) (1990), 167–172.

Killing, J. P. *Strategies for Joint Venture Success.* Praegler, New York, 1983.

Korbmacher, E.-M. Organizational problems in supra-company project management. *Nordnet 91 Conference.* Trondheim, 1991.

Lee, M. K. and Lee, M. K. High technology consortia. *High Technology Law Journal*, **6**(2) 1991, 335–362.

Lloyd-Schut, W. S. M. Construction joint ventures: EEC competition aspects. *International Construction Law Review*, **9**(part 1) (1992), 3–10.

Loraine, R. K. *Construction Management in Developing Countries.* Thomas Telford, London, 1991.

Merna, A. and Smith, N. J. *Guide to the Preparation and Evaluation of BOOT Project Tenders.* Project Management Group, University of Manchester Institute of Science and Technology, Manchester, 1993.

Morris, P. W. G. and Hough, G. H. *The Anatomy of Major Projects.* Wiley, Chichester, 1987.

Pfeffer, J. and Nowak, P. Joint ventures and interorganizational interdependence. *Administrative Science Quarterly*, **21** (1976), 399–418.

Schwartz, E. A. Disputes between joint ventures: a case study. *International Construction Law Review*, **3** (1986), 360–374.

Smith, N. J. and Wearne, S. H. *Construction Contract Arrangements in EC Countries.* European Construction Institute, Loughborough, 1993.

Swierczek, F. W. Culture and conflict in joint ventures in Asia. *International Journal of Project Management*, **12**(1) (1994), 39–47.

Walmsley, J. *Handbook of Joint Ventures.* Graham and Trotman, London, 1982.

Williams, R. G. and Lilley, M. M. Partner selection for joint-venture agreements. *International Journal of Project Management*, **11**(4) (1994), 233–238.

CHAPTER TEN

Procurement strategies for privately financed projects

N. Smith

Introduction

The purpose of this chapter is to demonstrate the fundamental roles adopted by the main stakeholders under their respective contractual arrangements in situations where private sector finance is being used to wholly or partially fund a project. Two main types of organisational structure are examined and the implications for all aspects of procurement reviewed. An investigation into the increasingly important use of public–private partnerships (PPPs) is presented, and the chapter concludes with an overview of recent trends and developments.

For more detailed information regarding all aspects of private finance rather than just procurement, see Merna and Njiru (2002).

General procurement principles

As stated clearly in the earlier chapters of this book, 'procurement' is the term used to describe the overarching process of the identification, selection and acquisition of civil engineering services and materials; their transport, their execution or implementation; and subsequent project performance. It includes the 'internal' aspects of administration, management, financing of and repayment for these activities.

With the more traditional forms of infrastructure procurement, where the state uses public money, from taxation or borrowing, to provide the capital costs and in some cases also the operational costs, it is often difficult to identify some of these procurement routes. In contrast, in privately financed projects the traditional roles and responsibilities of the client have been taken on by the promoter organisation and its contractual partners, and each aspect is clearly defined and regulated.

The life cycle of a typical construction project is shown clearly in Figure 1.2 (see p. 7); superimposed on the cycle is the range of procurement options over time. The privately financed option shows that the promoter takes responsibility for all aspects of the project from some point during the appraisal stage of the life cycle until a predetermined date during the operating stage. Within a concession type of agreement the promoter is established as a special project vehicle (SPV), and as a 'man of straw' that is a legal entity with no real asset value. Hence all services and materials will have to be procured through secondary contracts.

Privately financed projects are illustrative of two key principles, privatisation and collaborative working. Both of these have direct implications for procurement. Privatisation obviously indicates the procurement of funds, the payment of fees, the generation of income utilising entrepreneurial expertise, and some transfer of risk from the public to the private sector. Collaborative working operates at two main levels: at the SPV level, where individual staff from different organisations are brought together, usually for the duration of the concession, but ultimately returning to their original employer; and at the secondary-contracts level, where external organisations have contracts for parts of the concession period with the promoter organisation. A simple organisational structure for this type of arrangement is shown in Figure 10.1.

Concession contracts

The definitions and terminology associated with private finance are becoming more firmly established, although usage is not universally consistent. The use of private finance for major 'public sector' projects is not new. Many private canal and railway transport projects were undertaken in England in the 1880s and 1890s requiring approval by Act of Parliament, prior to shares being issued to individual and corporate private investors to raise the capital. An engineer would then be engaged to procure the necessary goods and services to implement the works. The private owners would then operate the service and charge the public and other private users. Better-known international examples include the Suez and Panama Canals and the Trans-Siberian Railway.

These types of project tended to be overshadowed by state- or public-sector-funded projects in most countries during the 20th century, but by the mid-1980s pressures on public debt and levels of taxation revived interest in the concept. Generally, the use of private sector funding for a typically public sector project has become known as a 'concession contract'. Such contracts are also known as

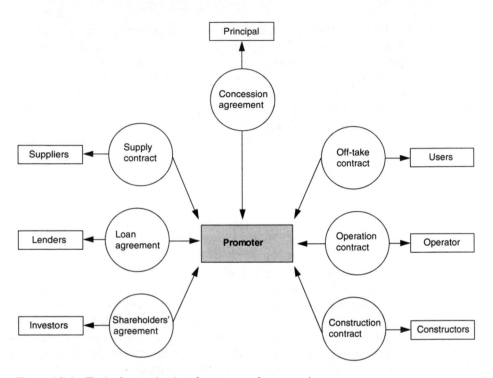

Figure 10.1. Typical organisational structure of a concession

build–operate–transfer (BOT) contracts, and many other acronyms exist for variants of these procurement strategies.

Concession contracts may be defined as a project based on the granting of a concession by a principal, usually a government, to a promoter, sometimes known as the concessionaire, who is responsible for the construction, financing, operation and maintenance of a facility over the period of the concession before finally transferring the facility, at no cost to the principal, a fully operational facility. During the concession period the promoter owns and operates the facility and collects revenues in order to repay the financing and investment costs, maintain and operate the facility and make a margin of profit.

The reason for the name 'concession contract' is easy to understand. Effectively, the client, usually the government, is awarding a concession to a private group, usually known as the 'promoter', to provide some predetermined type of service or operating facility of the type normally in the public sector for the period of the concession.

Secondary contracts

The term 'secondary contract' is given to each of the contracts initiated by the promoter organisation to undertake the project. It should be remembered that the promoter is usually an organisation specially set up for the concession with minimum asset value. Hence the need to use contractual arrangements to transfer work, risk and incentive to other parties. The most important secondary contract is the loan agreement drawn up between the promoter and the major provider of debt finance. Some terms are defined below:

- *Loan agreement*: the contract between the lender and the promoter. Lenders are often commercial banks, niche banks, pension funds or export credit agencies, who provide the loans in the form of debt to finance a particular facility. In most cases one lender will take the lead role for a lending consortium or a number of syndicated loans. In the event of default by the promoter the lender will have a contractual right to take over the concession agreement, complete the project and operate the project to return the investment.
- *Shareholder agreement*: if needed, this is a contract between investors and the promoter to purchase equity or provide goods in kind and forms part of the corporate structure. The shareholders may include suppliers, vendors, constructors, operators and major financial institutions, as well as private individual shareholders. Investors provide equity to finance the facility, the amount often determined by the debt/equity ratio required by lenders or by a provision of the concession agreement.
- *Operations contract*: between the operator and the promoter. Operators are often drawn from specialist companies or companies created specifically for the operation, maintenance and sometimes training requirements of one particular facility.
- *Construction contract*: between the constructor and the promoter. This may be a series of contracts for outline design, detailed design, construction and commissioning or a type of turnkey arrangement. Constructors are often drawn from individual turnkey or private-finance construction companies or a joint venture of specialist construction companies.

In 'market'-led concessions these secondary contracts form the basis for the project. As the term suggests, the promoter organisation is carrying the market risk. For toll roads or estuarial crossings where revenues are generated on the basis of directly payable tolls for the use of a facility, no off-take contract is required. Where a service or product is being produced, it may be possible to reduce project risk further by introducing two additional contractual agreements:

- *Supply contract*: between a supplier and the promoter to supply raw materials to the facility during the operation period. Suppliers are often a state-owned agency, a private company or a regulated monopoly.
- *Off-take contract*: in contract-led concession projects such as power generation plants, a sales or off-take contract is often entered into between the user and the promoter. The users are the organisations or individuals purchasing the off-take or using the facility itself.

Summary of the procurement procedure

The first phase of procurement requires the principal to determine whether a concession project strategy should be advocated after an initial feasibility study. If so, and it will not be so in every case, the next stage is the pre-qualification stage. This forms the basis for identifying suitable promoter organisations wishing to bid for a particular project. A draft concession agreement is prepared by the principal on the basis of the terms of the concession. A number of suitable promoter organisations are then selected. The concession agreement forms the basis of the principal's invitation to tender.

The second phase requires a number of promoter organisations to assess the commercial viability of the project based on the principal's requirements as identified in the concession agreement. A detailed appraisal indicates whether the project is commercially viable and whether the promoter should proceed with the bid. Secondary contracts are then identified and their influence on the project considered. If the promoter deems the project commercially viable on the basis of the secondary contracts, then a preferred bidder or bidders is/are identified.

In the final phase, the principal initially appraises the conformity of each bid based on the contract documents and then evaluates each bid according to the package weighting identified by the principal at invitation stage. The final concession agreement is then negotiated with the preferred bidder and a contract awarded. If agreement cannot be reached, the principal has to negotiate with the 'second' preferred bidder to reach agreement.

Invitation to pre-qualify

Concessions can either be sought or invited by the principal or be speculative ventures proposed by a promoter organisation. In the

European Union, any project over 3.5 million euros has to go to competitive tender. This accounts for the overwhelming majority of concession contracts. However, around the world about 60% of concession projects are speculative.

In the UK, principals invite promoter organisations to pre-qualify. Unlike conventional methods of pre-qualification, which may consider contractors for a number of future projects, pre-qualification for a concession project is usually particular to a project. However, in the case of small-scale 'bundled' or 'portfolio' concessions now being considered in the health and education sectors, the successful tender will be given future work subject to satisfactory performance. It is important to note that the procurement process has evolved since the early 1980s and is continuing to improve as more concessions are agreed. The invitation to tender would normally outline the scope of the project, its location, the programme proposed, the concession period and, if applicable, the required tender bonds.

Pre-qualification of promoter organisations may include assessment of a financial report of the promoter SPV. This may include the binding joint venture contract and any proposed secondary contractors, the proposed sources of finance, the debt/equity ratio, past concession work carried out by the promoter, and the range of relevant technical, operational and financial capabilities necessary for the project, together with resources and references. Relevant health and safety records, quality assurance systems and the industrial relations history may also be requested. There is no standard pre-qualification document, and some principal organisations request data not required by others. The final selection of pre-qualified consortia is usually made on the basis of meetings with and presentations from the promoter SPVs.

Concession agreement

In all concession contracts, it is the concession agreement which defines the relationship between the principal and the promoter SPV. A concession agreement consists of two distinct sections, first the legal agreement comprising the general, specific and common terms, and second the project conditions comprising the construction, operation and maintenance, finance, and revenue generation packages. It identifies all risks, rewards and responsibilities of the parties. Typically, in the UK, it would be drafted by the principal or its advisers and be issued at tender with the intention of forming the basis for final negotiation of the contract between the principal and promoter. The terms and conditions are as follows:

- *General terms* set out the nature of the agreement. They would normally include items such as the purpose of the concession, the commencement date of the concession, the concession period, the procedures for granting the concession, the rights of ownership, and any matters relating to confidentiality and exploitation.
- *Specific terms* contain the details of how this particular agreement will operate and how many of the risks and obligations are shared. Naturally, many of the terms relate to financial and/or contractual issues. Normally these might include restrictions on the assignment of the concession, the terms of payment, the degree of commercial freedom in operation, procedures for making adjustments to the concession, exclusivity, compensation rights, taxes and duties, how to deal with any existing facilities covered by the concession, interest rate guarantees, liaison procedures, quality assurance systems, tax incentives, and any specific government incentives and support available or required under the concession.
- *Common terms* are key factors affecting both parties, including changes in legislation, necessary insurances, procedures for termination, and procedures for both *force majeure* and dispute resolution.
- *Project conditions.* A concession contract is sometimes referred to as a 'four package' contract, where the four packages would be a technical package, an operational package, a financial package and a revenue generation package. In order for the promoter to be able to prepare this 'package' information, the concession agreement has to clearly indicate the standards and constraints affecting each package:
    - *Technical:* required design standards and specifications, design life, the layout of existing services, the maximum construction period, the method of bringing into operation or commissioning, any restrictions on the source of materials or on the method of construction, programme of related works if any, and warranties required during construction.
    - *Operational:* a performance specification, minimum demand levels and/or capacities, the transfer method at the end of the concession, testing procedures, methods of measuring off-take, accounts and records to be maintained, equipment inspection procedures, vendor-operated equipment, 'shutdowns', and staff training if applicable.
    - *Financial:* debt/equity ratio, coverage ratio, working capital, dividends, standby loan facility, shareholder agreements if any, currencies of loans, and sources of finance.

    –   *Revenue generation:* levels of tolls or tariff, period over which tolls or tariff will be levied, frequency and method of toll or tariff adjustment, minimum demand guarantee, escrow arrangements, guarantee on minimum demands, and subsidies.

In the UK, the principal prepares the tender documents, including a draft concession agreement, the instructions to tenderers and the criteria for award. The criteria for award indicate to the bidders how the returned tenders will be assessed. This ensures that all bids will be comparable and simplifies the tender evaluation process. Typically, the criteria might include confirmation of meeting the terms of the concession, information regarding the relative weighting of each project package and other factors specific to the project.

When evaluating bids, the principal must first establish compliance with the invitation to tender, with the general, specific and common terms of the concession, and with the project conditions. It is rarely sufficient to use price as the sole criterion for evaluation, and a number of factors will have normally been included as criteria for award. Any evaluation process must also include the managerial quality of the bid, its effectiveness and efficiencies, and cost savings to the user and also to the principal upon transfer.

A matrix or criteria-weighting system associated with each bid could be adopted. If four bids were submitted then the principal would initially check that each bid met the terms of the concession and the project conditions identified in the concession agreement. The principal would then evaluate and compare the components of each package of each bid based on the weighting system. The principal would award each package a specific number of points based on the evaluation.

Public–private partnerships

The first generation of concession contracts were large but profitable projects in which the private sector was responsible for raising all of the necessary finance. This is not to say that the concession is based solely on an alternative source of finance; it also utilises the efficiencies and entrepreneurial skills of the private sector to generate revenue. Hence the main general reason that governments are now looking closely at concessions includes the fact that public infrastructure and services can be provided more efficiently if the private sector is involved. However, many of the current generation of concession projects are politically and socially attractive to the principal but are not sufficiently financially robust to be fully funded by the private sector; hence a balance of public and private sector financing is required. These types of concession are known as public–private partnerships.

The private sector is permanently looking for opportunities for profitable investment. In recent decades, public infrastructure has not attracted significant quantities of private funds. This is partly because of the very long-term nature and therefore the long payback period of the investments, and because of the problems of charging users, but also because of the traditional central role of the public sector in most countries. In the UK, it is only since the Bates Reviews of 1997 and 1999 that private sector involvement in the full range of public sector infrastructure began to be realised.

It has been stated by private organisations that a main obstacle to greater involvement in PPPs was that public sector procedures and decision-making were too bureaucratic and unwieldy. At the same time, public sector representatives felt that the private sector was looking for unrealistic rates of return and mitigation of risks. More-over, public sector officials feared that handing over projects to the private sector would leave them without control of the project but with the responsibility for it. A further complication arises because public investment decisions are usually open democratic processes, including full details of the sums of money involved, while private decision-making is a more closed, confidential proce-dure; this can complicate negotiations between the sectors. A lack of understanding of the respective roles and requirements of poten-tial public and private partners could be an obstacle to developing PPPs.

PPPs can involve a private sector contribution from 0% to 100%; that is, they may vary from simple commercialisation to complete privatisation. There is no single model, and each scheme has to be tailored to the particular circumstances. Structures are very often quite complex. Nevertheless, there are some clear common elements to PPPs. Great care has to be taken at the start in setting up the right structures so that the roles and responsibilities of the partners are clear and agreed. Conflicts of interest should be identified and prop-erly regulated. The project needs a strong owner. The sharing of risks and responsibilities has to be negotiated in detail. The general prin-ciple is that risks should be allocated to those who can best control them. Risks which have been mitigated and managed should then be used to re-examine the financing arrangements. This is an iterative process at the early stages of the project. There is a major step change at commissioning, where the project moves from an implementation phase into operation, which is widely perceived as less risky. At this point there is an opportunity to refinance the concession with money borrowed at lower rates of interest.

The structure of contractual relationships and the contracts themselves play a crucial role in concession contracts. The principal plays a complex role, trying to balance the need to regulate to ensure that public accountability requirements are satisfied with the need to be flexible to allow the private sector to operate effectively. A key lesson from experience is that contracts need to be more flexible, specifying mechanisms more than details. It is not possible to put everything into the contract and it is pointless to try. The emphasis should be on output specifications rather than input specifications, setting out clear performance criteria rather than predetermined technical requirements. This approach allows the private sector to innovate while ensuring that public sector requirements are met. A pragmatic approach is needed. Concession contracts usually cover long time periods and it is impossible to make reliable forecasts for 30 or 40 years, therefore the contracts should contain clear mechanisms for their amendment over time.

If a project is considered to be viable, i.e. the identified risks appear to be commensurate with the investment to be made and the potential of realising a commercial return, it could be financed solely from equity and debt finance from the private sector. However, if the project is assessed to be too risky, it may either be suitable for PPP or be a project not suitable for the concession type of contract. In these types of arrangement both the public and the private sectors have a role to play in project finance and the allocation of risk, as presented in Figure 10.2. There are usually two ways to allocate risk: through a payment mechanism where the basic debt and equity instruments can be combined to maximise the project's cash flow, and through specific contract terms. This manner in which the public sector contributes to the PPP is described in detail in the following section.

Public sector finance

Figure 10.2 is a schematic illustration of the possible options that the public sector can take to help attract private investment, and to minimise the private-sector-financing cost premiums. The private sector debt and equity funding can be supplemented in a number of ways ranging from an additional sharing of risk in the concession agreement, a relatively easy option, to the injection of direct finance, usually the last resort.

Risks associated with PPPs relate to both the commercial project and the risk of the project financing itself. Financial risks are

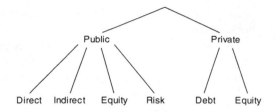

Figure 10.2. Spectrum of public and private finance

associated with the mechanisms of raising the delivery of finance and the availability of adequate working capital, and can be seen as largely proportional to the riskiness of the project. In the concession contract risks should be equitably divided between the public and private parties involved. Certain risks are better managed by the public sector. A risk which neither party can control should be placed with the public sector because action will only be required if the risk should occur, whereas the private sector will always put a price on risks it cannot control, which will be 'paid for' in terms of the cost of services and affect value for money whether the risk occurs or not.

The types of risks to be borne by the public sector in its role as principal in concession PPPs will be discussed below. However, as a general rule, the principal should be prepared to retain some or all of the risks where this does not threaten the incentive for efficiency gains by the private sector, where the risk is largely outside the control of the private sector and where the risk can only be transferred at a cost to the private sector which is far higher than retaining the risk in the public sector.

Typically, in the concession agreement, the principal carries the responsibility for the global risks. These normally include risks outside the project elements but which can influence the performance of the concession. They include the political, legal, commercial and environmental risks affecting the concession. These risks, being outside the management and control of the promoter, cannot be allocated through secondary contracts, and are therefore usually retained by the principal. The public sector is likely to resist such a contractual position, claiming that legislative risks are part of everyday business. As such, these risks will probably have to be much debated and a negotiated position reached during the pre-completion stage of the agreement.

In a PPP the principal may be required to take responsibility for additional risk so as to make the project acceptable to the private sector. One obvious risk is the demand/revenue risk, which can be borne in a number of ways, for instance by providing a revenue support guarantee in case of traffic flows falling below an agreed level, or by ensuring that new developments will not be directly competitive with and thus affect the revenue stream of a facility. The public sector can also enter into commitments to replace direct charges by proxies such as shadow tolls or leaseback fees.

Equity provision

Equity provision by the public sector is the third option shown in Figure 10.2 and would be used if the redistribution of risk alone was not capable of making the concession attractive. Equity from the principal is usually in the form of land or, quite often, in the form of an existing facility with an existing revenue-generation stream, which is transferred to the promoter. The operation of an existing facility can provide an immediate income, which the promoter can use to reduce loans and repay lenders and investors early in the project cycle.

Public equity gives the public and private sectors a common interest in the project and helps to raise loan finance by improving debt/equity ratios. It can also be sold once the project has a proven track record, through a share flotation or a private placement with institutional investors. Public or institutional investors can invest in equity shares through a shareholders' agreement as discussed earlier in the chapter. Sometimes, third parties such as the European Investment Fund may take equity shares in selected projects which satisfy their investment criteria.

Indirect finance

If further action by the principal is needed to make the PPP viable, indirect methods of financing may be considered. By providing indirect forms of finance, principals can exert influence on the financial attractiveness of PPP projects by reducing their financial risks without investing directly. This is achieved through the use of numerous financial instruments, which include taxation procedures and financial measures.

One of the common indirect tax procedures is the use of a 'tax holiday'. Here, the principal grants the promoter a tax exemption period following completion date. In the UK concessions are 'ring-fenced' from the viewpoint of taxation but, once profitable, become liable for corporation tax. By granting the tax break, the principal is letting additional monies be retained by the promoter at a time when repayments will be high. The principal does not receive any income from tax during the period of the break, but after the break period tax will be paid from a facility which could not have been completed without the indirect assistance.

Tax reduction/corporation tax advantages where the promoter is given a reduced rate for a period following completion may also be used. Again the principal does not gain the full rate of tax, but a lower rate of tax from an operational facility is of more value than insisting upon the full rate and having no facility to tax.

Governments can borrow money more cheaply than can the private sector. Therefore the other major indirect financing approach is to is to increase the 'bankability' of a project by providing credit enhancement and/or extending the loan repayment period, thereby reducing the financing costs. Structurally subordinated loans, which are loans with longer maturity periods (over 20 years), including the use of bullet loans with a single capital repayment on maturity, and extended grace periods for capital repayment are possible options.

Debt finance

This option from Figure 10.2 is very much used as a final resort. However, in the case of a risky project a debt commitment with a guaranteed loan from the principal may be the only way of reassuring the private sector of the viability of the concession. When banks are asked to participate in the debt financing of a project they look at the percentage of guaranteed loans. The higher the ratio of guaranteed loans to normal loans, the more likely banks are to be attracted to the financing of the project.

Early operational-stage loans are short-term public sector credits which can be sold on to the private sector once the project has reached financial stability, allowing public sector resources to be recycled into new projects via a revolving fund mechanism. This type of fund has been proposed in the USA under legislation passed by President Clinton.

Public sector comparator

There is another factor in the procurement of concession contracts that is not present in other types of procurement, and that is the public sector comparator (PSC). In any decision about the use of public-only finance or public–private finance of projects the principal ultimately has to establish the value of the concession to society. Principals need to consider the price of risk transfer, by asking whether the amount of additional risk taken on by the private sector exactly equals the amount of risk reduced from the public sector. However, it should be noted that despite the benefits achieved from PPPs, it is argued that the proportion of the overall financing provided by the private sector should be kept to the lowest level possible because of the higher risk premiums when compared with publicly subsidised or publicly backed debt.

To this end, a number of measures have been developed to measure the cost of conducting a project in the public sector, compared with the cost of using private sector money, for example the value-for-money test developed by the UK Treasury in 1998. However, the main UK method for evaluation and choice of public-only or public–private finance is the PSC, which enables the government to compare concrete PPP tenders with traditional, purely public alternatives.

Procurement of PPP

Figure 10.3 illustrates a schematic overview of the major phases of the PPP process. These phases are further discussed and explained in the following text.

Initially, the suitability of the concession to operate as a PPP has to be established. Support for PPPs by the principal is a major factor and without full support it would be difficult to proceed. It should always be remembered that PPP concessions are not a panacea and that some projects might be more suitable for the public sector domain. There should be consideration of the factors that will ensure the effective application of a PPP: technical and organisational, marketing and financial, legal and administrative, policy and regulatory, and political. Is there sufficient flexibility granted to the promoter organisations to enhance the project's commercial viability to improve project viability? If not, the PPP option should be abandoned.

Once the basic features of the PPP have been agreed, two types of analysis should be undertaken for risk and finance. A financial analysis and a socio-economic analysis are usually completed so as to determine the extent of the PPP funding 'gap' in the project. This information will indicate whether the project's objectives meet the required PPP objectives; this will then indicate whether the project can proceed to tender stage, whether it should be abandoned or whether it needs to be re-engineered. Sometimes the restructuring of projects or the combination of project ideas into group or portfolio projects can produce stronger PPP concessions.

At tender stage the project criteria are passed from the principal to potential promoters, who should prepare a pre-qualification

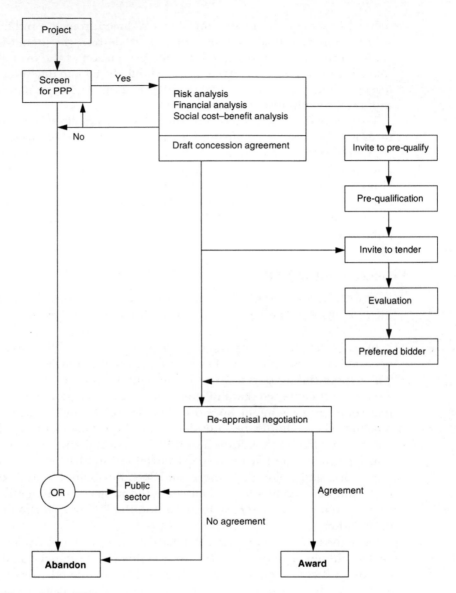

Figure 10.3. PPP procurement

document for evaluation. Potential tenderers will be shortlisted and supplied with a draft concession agreement outlining the terms and conditions of the contract. On the basis of this, selected tenderers will submit their bids for evaluation by the principal; the preferred tenderer will then be identified.

Conclusion

The procurement of concession contracts and PPPs can take many forms, from large, international projects to small, bundled, social projects. The restrictions on the use of public funds combined with the continued demand for infrastructure means that the procurement of such concessions is likely to be of significance in the short-term and possibly medium-term future. Although of a different culture from the traditional public sector procurement methods, concessions offer viable alternatives for long-term agreements between the public and private sectors to provide and operate infrastructure and/or services. PPPs involve a sharing of responsibility and risk by the public and private partners.

PPPs have the potential to provide, through financial engineering, an optimum combination of public sector and private sector approaches. However, PPPs are not a panacea and are not the best solution in many cases. The procurement of concessions and concession PPPs has to be structured with great care at the start to make the project requirements, roles and responsibilities clear and to regulate conflicts of interest. Contracts should cover all the major aspects of the PPP project but should allow some flexibility for innovation and economy, specifying performance requirements and not necessarily technical details.

Bibliography

Carter, B., Hancock, T., Morin, J.-M. and Robins, M. *Introducing RISKMAN Methodology*. The European Project Risk Management Methodology. Blackwell, Oxford, 1994.

European Commission. *PROFIT 2001. Private Operation and Financing of Trans-European Networks – Public Private Partnerships*. DG TREN 5th Framework RTD project ST-98-SC-3035. European Commission, Brussels, 2001.

Levy, S. M. *Build, Operate, Transfer: Paving the Way for Tomorrow's Infrastructure*. Wiley, New York, 1996.

Merna, T. and Dubey, *Financial Engineering in the Procurement of Projects*. Asia Law and Practice, Hong Kong, 1998.

Merna, T. and Njiru, C. *Funding and Managing Infrastructure Projects*, Asia Law and Practice, Hong Kong, 1999.

Merna, T. and Njiru, C. *Financing Infrastructure Projects*. Thomas Telford, London, 2002.

Merna, T. and Owen, G. *Understanding the Private Finance Initiative*. Asia, Law and Practice, Hong Kong, 1998.

Merna, T. and Smith, N. J. *Guide to the Preparation and Evaluation of BOOT Project Tenders*, 2nd edition. Asia Law and Practice, Hong Kong, 1996

Merna, T. and Smith, N. J. *Projects Procured by Privately Financed Concession Contracts*. Asia Law and Practice, Hong Kong, 1996.

Merna, T., Payne, H. and Smith, N. J. Benefits of a structured concession agreement for build–own–operate–transfer projects. *International Construction Law Review*, Jan. (1993), 32–42.

Smith, N. J. (ed.). *Managing Risk in Construction Projects*. Blackwell Science, Oxford, 1999.

CHAPTER ELEVEN

Framework agreements

D. Bower and P. Garthwaite

Introduction

As clients recognise the value of long-term arrangements, framework agreements are becoming increasingly popular. A framework agreement in itself gives no work to the contractor and may be non-exclusive. It is a long-term commitment between the parties that enables clients to place contracts on pre-agreed terms, specifications, rates, prices and mark-up that are embedded in the framework to cover a certain type of work over a period of time or in a certain location or both.

The contractor makes staff, designers and construction resources available to undertake these contract packages as they are awarded and ensures their completion within agreed standards and timescales. The framework agreement is developed to deal with a variety of issues and these are explored in this chapter, including setting up the strategic alliance. In this chapter the term 'alliance' is used to cover all forms of collaborative arrangement between parties. The establishment of a framework agreement and the elements requiring consideration are detailed, and a framework model is described.

Partner selection

Many companies recognise that by working closely with other organisations they will be able to develop, produce and sell their products more successfully over an extended period of time. Combined efforts can improve a firm's alien appearance when entering a new market locally, nationally and internationally; they can optimise risk sharing; and they can develop knowledge of new technologies. In essence, such alliances are formed when the perceived benefits outweigh the expected extra costs. When one is forming an alliance, the elements shown in Figure 11.1 must be considered. These are referred to as the 'enabling elements'.

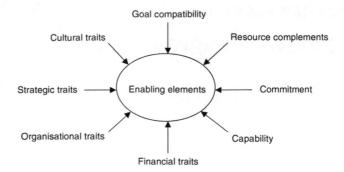

Figure 11.1. Enabling elements in alliance formation

Cultural traits

Collaboration and cooperation between parties from differing cultures will always create complexities within any situation. Friction can occur, firstly, when one party unilaterally imposes its own culture and normal behavioural standards on another party within the agreement and does not consider the cultural attributes that could be brought by the other party. Secondly, one party may inadvertently give up its own culture to the other party.

Corporate cultures also have an important influence on partner selection. Potential partners need to assess how well differences between organisational cultures can be managed to develop a mutual trust. This is done by looking at the organisational and management compatibilities by asking questions such as: What differences exist in organisation structure or business strategies between the parties? Is decision-making centralised or decentralised? How compatible are the company visions and mission statements? Are both management teams committed to overcoming cultural differences?

Strategic traits

'Interpartner fit' is a term widely used to describe the skills that already exist in the partnering parties. This is a party's ability to acquire, copy, integrate and manipulate knowledge and skills, and depends significantly on how newly acquired knowledge links fit with their existing knowledge bases. This is often called a party's absorptive capability, and it is from these knowledge bases and links that the strategic organisation fit is determined. A party's absorptive capability is positively correlated with an alliance's profitability and sales growth.

Industrial and business background, market position, and marketing and distribution networks are a good indication of a party's market power. Should a party's market power be proportionally significant it can result in the party being able to direct industry output, increase its bargaining power and also offer economies of scale. This can be a major advantage in a multi-project environment where bargaining is required for numerous projects, and also regarding the economies of scale when considering a multi-project environment of the kind described in Chapter 13. Strong market powers can also result in greater bargaining power with governments, and thus in reducing the political risks and business uncertainties.

Linked with market power is market experience. An established history and strong background in a market generally mean that the party will have a solid marketing and distribution network. Business activities by the strategic alliance could be easier to integrate into markets using parents' connections and relationship networks.

Product relatedness between the parties can be helpful in the establishment of a long-term relationship between parties. Should parties create similar products before the formation of the framework agreement, it could result in an economy of scale and influence the scope and efficiency of transaction costs owing to the use of existing distribution channels, production facilities, marketing skills and consumer loyalty. This could also help to establish relationships outside the framework agreement with suppliers, distributors, customers and governments.

Organisational traits
Geringer (1988) pointed out that economies of scale, market power, process innovation and organisational image can all be linked to a party's organisation size, and that organisation size is positively linked to any strategic alliance's survival and growth. Alongside the above, a larger organisation size allows an improved ability to reduce risks and mitigate uncertainty. One concern with large conglomerate parents is that strategic alliances do not always receive the strategic attention and commitment required.

International business experience is always attractive to potential partners because it represents wider perspectives than those of others, an advanced knowledge of other jurisdictions and possible adaptability and willingness to cooperate. Previous work between the potential partners is also an advantageous trait when considering framework agreements. Economic transactions between parties help to improve social-relation embedment between the partners, and thus help instil trust and deter opportunism within the framework agreement. Secondly, previous successful cooperation between the parties can lead to the development of skills and routines that are specific to any following relationships between the two parties and can be used, on a form of trust within a framework agreement, to reduce the number of control measures (discussed later). The operation and management of joint values can be improved if the partners have understood and correctly acknowledge each other's strengths and weaknesses.

Other organisational attributes that need to be considered include organisation skills such as the ability to blend the various techniques and capabilities of job design, recruiting and staffing, training, performance appraisal, compensation and benefits, career development, and labour management relations. Not all the aforementioned are relevant to a framework agreement individually, but a number will be included within the consideration of the contributions to be made by the individual parties to the strategic alliance.

Financial traits

Financial interpartner fit concerns the cash flow position and capital structure similarities between parties. The idea is that the partners will be able to judge the total risk of a capital project and develop the appropriate risk-adjusted discount rates used when assessing the project. This will be a standard requirement in a multi-project environment owing to the anticipated use of the above systems and their importance.

Risk management aims to reduce the vulnerability of the venture to external hazards and internal instability. Flexibility may be required from all parties to allow for alterations to asset structures or even the whole venture or alliance, through an agreed variation to the framework agreement. This can result from numerous factors, including exchange rates, which are extremely important to international ventures.

Finally, and possibly most important of all, there is the financial ability of potential partners. There are a number of financial capabilities to be researched before a framework agreement can be developed. These are: profit making, including the ability to exercise cost control, increase revenue, reduce taxes and expenses, and maximise operational efficiency; allocation and utilisation of capital, including the ability to allocate and use working capital, obtain local financing, use and control debts, and manage risks; and assets management, including the ability to optimally deploy assets and resources, manage accounts receivable and cash flows, and manage fixed and intangible assets.

Goal compatibility

The compatibility of the goals set for a strategic alliance by its parents affects how the parents behave cooperatively or opportunistically. Different goals between the parent parties regarding the alliance or joint venture 'plant the seeds for subsequent opportunism and conflict'. Goal congruency reduces a party's uncertainty about the future behaviour of another party to the agreement, allowing a better response to the other party's strategies, and creating a better organisation fit and strategic symmetry between the parties. This goal congruency can stimulate the commitment by all parent parties, increasing motivation, and allows the alliance to move in the same direction collectively.

Goal congruency is critical in contracts of high complexity and a proposed long lifespan. Each parent needs to evaluate the objectives of potential partners, identifying their compatibility and concluding whether the parties could work together. If the answer is no, then consideration should cease at this point. Goal congruency harmonises the interests between parties that would otherwise give way to antagonistic and opportunistic pursuits.

It is important to not confuse goal compatibility with goal identity. It has been stated that the strongest environment for an alliance is one in which strategic goals converge and competitive goals diverge. So long as the partners understand and respect the other partner's goals, and the goals are not in conflict, a strong goal congruency can be created. An example is when one party is interested in the local market, whilst another partner is concerned with exports.

Complementary resources

Having complementary resources allows a reduction in governance and coordination costs, stimulating information exchange during diversification. It not only improves operational and financial synergies but helps to improve influential learning curves. Following from this it could be said that a framework agreement is a contractual and strategic method to combine complementary assets and resources, and aims to develop new skills that the partners lack to fulfil their strategic objectives, from the other partner(s).

The question of what are complementary resources and their rank within the framework agreement can be left to the individual situation. These complementarities can include missions, resources and managerial capabilities, or be complementary skills that create a balanced bargaining power and strategic fit between those privy to the agreement. Where each partner contributes one or more distinct elements in production or distribution, two examples of the possible use of complementary resources are vertical quasi-integration, and horizontal linkages between the parties' strengths in geographical areas.

Commitment

Some would say commitment is the most important element of the partner selection process, because without full commitment, a percentage of opportunism will exist. If partners are perfectly compatible, then without commitment to the framework agreement, when troubled times arrive, such as a change in market conditions, the individual parties will not be prepared to afford the time, finance or resources to keep the agreement functioning, and to strive to realise the strategic objectives.

Without commitment to the agreement, trust between partners will disintegrate, sowing the seeds for opportunism and frustrating one of the most important aspects of the agreement, cooperation. This will then lead to poor maintenance of the ties between the partners due to the breakdown of compatible goals, and the balance of risks will change.

Therefore, it can be seen that party commitment to the framework agreement is essential as a stabilising device against unexpected environmental changes and market developments. If commitment is high from all parties concerned, then interpartner conflict will not greatly affect the stability of the agreement, but if low, this could result in the primary source of instability being conflict between the parties.

Capability

Capability in this sense is not about having the ability to provide the resources or skills that are missing from the other partners. Here capability is the organisational capability of providing an essential supply base for the resources needed to create a long-term agreement.

Partners, through their due diligence studies, should use value-creating logic to assess the ability potential partners hold to tilt the market and competitive balance in favour of the strategic alliance. Hamel *et al.* (1989) conclude that three unique aspects provide capability: unique capabilities that cannot be trusted easily across companies; unique capabilities that cannot be easily substituted; and unique capabilities that cannot be independently developed or replicated within a reasonable time-frame. Examples include close interpersonal relationships with the local government authorities or the business community.

Confidence in partner cooperation framework agreements

So far in this chapter, the enabling elements of alliance formation have been discussed. From these it can be seen that within framework agreements there is an underlying potential for opportunistic behaviour by any partner involved, and thus an adequate level of confidence is required. framework agreements are interpartner cooperation plans to coordinate two or more partners pursuing shared objectives, and therefore satisfactory cooperation is fundamental. From this it is plain to see that a low level of confidence will lead to partners being suspicious of the other partner or partners (if a framework agreement is created in the first place) and the working relationship will deteriorate. The trust and control aspects of the framework agreement environment are the sources of confidence.

Partner cooperation and confidence in partner cooperation

A number of authors, when considering partner cooperation, have discussed the interpartner cooperation uncertainty that can exist. The definition adopted for this chapter is *the willingness of the partners to pursue mutually compatible interests that creates the framework agreement, rather than self-interest seeking with guile.*

Partner cooperation is characterised by honest dealing, commitment, fair play and complying with agreements. Truthfulness and commitment have different strategic effects within a framework agreement, where truthfulness is linked to developing confidence and commitment is an elemental building block of the agreement itself. Although the parties are expected to pursue their own interests, at the same time these must ensure that the framework agreement and the alliance contained within are followed correctly. This appears paradoxical, and a number of authors have gone on to state that if competition and cooperation are at variance, a sufficient level of cooperation cannot be achieved. Therefore, it can be concluded that partner cooperation is essential for a successful framework agreement to exist, but is difficult to achieve.

It is this prerequisite of partner cooperation that enables a confidence in partners and their cooperation to be built, but by no means is it as automatic as it sounds. Das and Teng (1998) define confidence in partner cooperation as 'a firm's perceived certainty about satisfactory partner cooperation'. This definition is quite vague and does not show a direction for the perception. Therefore, for this chapter, the following definition of confidence in partner cooperation will be used: *confidence in a partner's cooperation is the expectation of a partner's conduct regardless of the first partner's own.* This implies that the perception and judgement of confidence can only be away from the assessing party and have to be an evaluation of a fellow partner within the framework agreement.

Using the word 'expectation' helps to build a sense of uncertainty and implies a probability of an event occurring. It is here that the concept of control enters the equation. Probabilities imply risks and risk has to be managed and controlled. By controlling the risks, the confidence in the outcome of the framework agreement and the behaviour of the partners can be increased. Having now developed the existence of trust and control within confidence levels of partner cooperation, we shall now consider them in turn and examine the relationships they hold.

When a company has adequate control over its partners within a framework agreement, that company will have a reasonable confidence in cooperation from those partners. Das and Teng (1998) detailed a number of approaches to control and the terminology that is utilised. They explain that, due to diverse opinion, control can be considered simultaneously as an organisational set-up, a process of regulating behaviour and an organisational outcome. For use within this chapter, control is defined as *a means of operation to implement standards developed for the achievement of the framework agreement and partners' objectives.*

Control mechanisms and the level of control are two integral instruments to gain and keep control. Control mechanisms exist to help the achievement of a satisfactory control level, where the control level is the degree to which a party feels others are cooperating. From this it can be seen that control is used to improve the predictability of attaining the organisation's goals, creating more certain results and thus generating a confidence aura.

Partners over time can develop close bonds within framework agreements, forming positive feelings regarding each others' conduct, and it must be noted here that a minimum level of interpartner trust is required for a framework agreement to be formed and function. Every project in a multi-project environment contains an element of trust of all partners within the framework agreement. Should this trust be damaged in one project, a detrimental effect will occur in others within the framework agreement and the multi-project environment to differing levels.

Control mechanisms are organisational arrangements intended to enhance the level of control. On the one hand, some believe that control mechanisms undermine the trust level in framework agreements by implying that one party does not trust the other(s). This could possibly then result in a vicious cycle of 'if you do not trust me, I do not trust you.' In contrast, control mechanisms could be viewed as building mutual trust through the provision of a track record, developing trust from the successful results of previous outcomes. This is idealistic for a multi-project environment, where, after the completion of each individual project successfully, the level of trust will increase. Legalisation in terms of reliance on formal rules and standardised procedures can facilitate the development, diffusion and constructive institutionalisation of trust in organisational settings. It makes sense that a framework agreement should include within its structure a system of developing trust through control mechanisms.

Outcome control and behaviour (process) control are linked together under the title of 'formal control' and are defined as the employment of rules, goals, procedures and regulations describing the expected outcome and behaviour. 'Social control' is the people aspect that is always important and is about achieving desirable behaviour through organisational values, norms and culture developed within the framework agreement.

Using formal control by setting the boundaries of the framework agreement using specific performance goals for output control and specific processes for behaviour control means that full autonomy for the agreement relationship cannot be achieved. This can lead to an opinion of mistrust and creates a negative relationship between formal control and trust level. Research has already begun and it has shown that not only can contractual safeguards create a lowering of trust levels, but so can poorly designed formal control mechanisms.

Social control uses the assumption that people determine their own behaviour, and this is achieved using the influence of shared goals, values and norms. Owing to the personal-determination factor of social control, interpersonal respect and less mistrust create a base for trust construction to commence. Social control, owing to its nature, takes time to evolve and develop because social control blossoms through socialisation, interaction and training. Therefore, from the above section it can be seen that use of formal control mechanisms can hinder the level of trust among the framework agreement partners. Also, the use of social control mechanisms intensifies the level of trust between the framework agreement parties.

There can be unexpected negative occurrences resulting from using control mechanisms and this affects control. These negative occurrences could include operating delays, gamesmanship, negative attitudes and behavioural changes, and it is these negativities that result in framework agreement partners not achieving complete effective control using control mechanisms.

To back up the suggestion that trust is a moderator between control mechanisms and the control level, Goold and Quinn (1990) have stated that 'trust is a prime prerequisite of effective control' because the utilisation of control mechanisms requires a certain level of trust. Therefore, without a minimum level of trust, a framework agreement's goals, rules and development of teamwork could be impossible. With this minimum level of trust between partners, there is a smaller chance of control mechanism failure because the partners understand each other better.

It can be seen that trust induces desirable behaviour in framework agreements and, although a good level of control requires trust, trust on its own cannot be attributable to control. Trust is a personal perception and does not influence the others' behaviour. If that was so, trust would be a control mechanism. Therefore, trust must be a moderator between control and control mechanisms. It has been shown that when trust is high, control mechanisms are more effective in providing a sufficient level of control, i.e. trust allows easier operation of control mechanisms.

Building trust in framework agreements

Now that we have seen that trust is a necessity for a control mechanism to be used effectively to create a satisfactory control level, discussion needs to centre on methods of building trust within the framework agreement. There have been several methods identified, as follows: trust from communication; trust from risk taking; trust from equity preservation; and trust from interpartner adaptation.

Trust from communication

Open and prompt communication is mandatory for any interaction to take place, certainly for trust-building relationships. Communication removes possible difficulties in the operation of a framework agreement by providing the infrastructure for cooperation and the ability to resolve conflicts and unfavourable situations. Due diligence is an integral part of the framework agreement process, and without communication this method of trust building would not be feasible. This method also works in reverse, where proactive information exchange shows openness and trustworthiness. The passing of unsolicited and sensitive information shows intimacy and goodwill, thus developing a supporting environment.

This helps to provide the moderator for social controls through the development of cultural norms and common values, by creating a familiar and level communication arena as a result of continued information exchange. When partners are interacting in a friendly and equal environment, trust will be automatically developed.

Trust from risk taking

Trust is closely linked with risk and risk taking, with many authors believing that trust and risk form a reciprocal relationship, i.e. trust leads to risk taking, and risk taking (if successful) gratifies a sense of trust. Creed and Miles (1996) state 'trust begets trust', and from this point it is felt that a good start procedure to boost trust among partners in a framework agreement is to signal one's commitment and trust through a significant level of non-recoverable investment at the start of the framework agreement.

This significant investment does not have to be in one single trans-action but can be split into small, regular commitments, for example incremental resource commitments. Methods of trust building such as this help to develop an opinion about the party through all peers and are probably the simplest form of trust building. Locate a partner that has a good reputation of being honest, fair and trustworthy.

Trust from equity preservation

Equity in the sense of this chapter means that the partner contrib-uting the most tangible and intangible resources to the framework agreement should get the most from it. It has also been shown that people within an organisation need measurement and equity to appreciate interactions.

The idea of equity of motivation states that people are preoccupied with maintaining a fair relationship rather than concentrating on the efficiency and productivity of the relationship. This leads us to say that unfair relationships can result in one party feeling someone is taking advantage and that party's confidence, and even commitment, will reduce, despite the framework agreement being successful.

Therefore, it can be seen that equity is an important source of trust within a framework agreement and lack of equity can damage mutual trust. It is also clear that where a level of trust already exists between partners, extended periods of inequity create tension and develop a strain within this existing trust level. This in turn suggests that when trust is being built, the benefits from the framework agreement need to be on an equitable basis.

Trust from interpartner adaptation

If one partner adapts to meet the needs of other partners this develops trust, and interpartner adaptation is the adjustment of a party's behav-iour to fit between the partnering parties or between the framework agreement (in this instance) and the environment. There is a need for all partners within the framework agreement to be as flexible and willing as feasible to allow divergence from the framework agreement, when necessary to carry out the adaptations to build trust.

A good example that is regularly quoted in texts is the host country of a joint venture imposing new laws prohibiting majority equity shares by foreign parties; then both parties must adapt. Such a situation, although possibly painful in economic or other terms, would earn much trust from the partner. From this it can be agreed that adaptation for the benefit of the framework agreement would earn trust from your partner(s).

Control mechanisms in framework agreements

Now that we have discussed the trust-building tactics that exist in framework agreements, it is appropriate to discuss the control mechanisms used to create and develop confidence in a partner's cooperation. A number of control mechanisms already exist, such as cybernetic regulations and information-processing devices. Two main forms of control exist: formal control and social control. It is important that the differences between them are understood because it is the utilisation of these that creates the relevant control mechanisms for framework agreements. These mechanisms are goal setting, structural specification and cultural blending.

Goal setting

Locke and Latham (1984) describe goal setting as the establishment of specific and challenging goals within organisations. Objectives management is seen as a suitable method for enhancing control and boosting performance, with it being used by nearly all companies in some shape or form. In framework agreements, the partners usually aim for a high degree of goal formalisation using short-term rather than long-term goals. Short-term goals are used because performance evaluation and feedback are monitored more frequently.

It is extremely important that the definition of the goals is included within the control mechanism, along with measurement systems and reinforcement methods. Clear objectives also help to direct the framework agreement, allowing the establishment of rules and regulations. Goals also specify what each partners' contribution is expected to be and provide a background to find any incongruent behaviour from partners.

Having described the effect of goal setting on formal control, we can also say that goal setting has an influential effect on social control. Participatory decision-making that is necessary between partners develops a control as the understanding of each other improves, creating norms and collective values to be used in the framework agreement. This means goal setting forms a consensus gradually, reducing the likelihood of agreed objectives being ignored. Therefore, it can be seen that goal setting is a good system of integrating formal and social control within the concept of confidence in partner cooperation.

Structural specification

Structural arrangements, rules and regulations are the innermost part of formal control. Because of the high degrees of goal incongruence and performance ambiguity, formal control is particularly relevant to framework agreements. For the sake of this model, '*ex ante* deterrents' and '*ex post* deterrents' are the most appropriate.

Ex ante deterrents are designed to minimise a partner's incentive for opportunism. Examples include non-recoverable investments that lose much of their value should the agreement fail, and help to create responsible behaviour as a result of the stake in the agreement each partner holds. This has resulted in the suggestion that equal equity ownership helps to reduce opportunism and is thus a more effective control mechanism.

Ex post deterrents are the structural safeguards against opportunism. These are reporting and checking devices, written notice of any departure from the framework agreement, cost control, quality control, arbitration clauses, and lawsuit provisions. The structural specifications are agreed during the negotiation stage, and if this does not occur it has been suggested that bargaining power is the next resort. In essence, rigid structural arrangements set the boundaries for the conduct of the framework agreement partners.

Cultural blending

As expected, an integral ingredient of social control is organisational culture, where organisational culture is defined as the shared values and norms that state the expected attitudes and behaviours from all partners. It is these shared values and norms that provide the control through the voluntary behaviour of partners to fit into the environment of the framework agreement.

The main concern with framework agreements for the partner organisations is loss of identity. Therefore, the challenge is to make cultural blending work, but ensuring that identities remain as individual as possible. Another challenge that exists is the blending of very differing cultures, and the prime example often used is a rigid, large company together with a flexible, small company.

Culture management is integral to success mainly because there are no alternatives, unlike goal setting and structural specification. The key to successful culture blending is socialisation, by providing interaction periods where managers familiarise themselves with the other partnering companies. This helps to develop the common values and norms for the framework agreement.

Ownership balance and structure

If the framework agreement relates to a specially formed company then framework agreement ownership has a strong effect on the risk sharing and resource commitment of the partners, and the vulnerability and strategic flexibility of the framework agreement. Here, framework agreement ownership means the balance of the privity of each partner relevant to its contribution or the division of the equity invested. If an agreement between two partners is equally split then this is an equal partnership. Any other balance will result in a majority–minority ownership. Both majority–minority and equal ownership can exist when multiple partners exist.

The balance of equity often reflects the profit remittance scheme, showing how any benefits or losses are distributed throughout the partners. The ownership balance/structure often shows a partner's investment strategy, owing to the close relationship between a company's capability to contribute strategic resources and the balance of ownership within the framework agreement.

The assets created for the benefit of a framework agreement, for example human capital, are often intangible and ownership-specific. Therefore, when the collecting of resources and core competencies occurs, it is the equity distribution within the framework agreement that allocates the responsibility for provision.

Finally, the exposure of a framework agreement to host government intervention can result in the development of a specific equity balance. As with any management forte, high interaction with the environment requires a need to decentralise decision-making power, i.e. a local company from the jurisdiction where the framework agreement is set would be best placed to take a majority control when there is high governmental intervention. This can be developed in the opposite direction by stating that when foreign partners hold a greater proportion of equity, lower risks and uncertainty are assumed by their effort towards the framework agreement.

The different ownership structures within a framework agreement

It is clear that for each partner, three types of ownership exist: (1) a majority ownership; (2) a minority ownership; and (3) an equal split or 50:50 ownership (if there are only two parties).

The main theory behind majority ownership is derived from transaction cost theory, where majority equity is necessary to gain dominance so that the party can effectively minimise transactional risks. Another advantage of a majority–minority ownership is that there is a reduced need for interpartner negotiations and bargaining during any decision-making processes.

It must be stated though, that majority ownership is not free. Well-documented examples are Xerox entering China and 3M entering China, where joint venture agreements were entered into in which Xerox and 3M were the majority holders, but had to endure a 50:50 profit split and export of all goods, respectively. It has also been noted that unequal balances can result in a change in the equilibrium of dependence, with the majority owner becoming more dependent on the partnership and therefore losing some of its relative bargaining power.

There is also a concern that the majority stakeholder may strongly push denouements to a vote, knowing it has a stronger influence. This can create a reduction in the commitment of the minority stakeholder, culminating in the removal of its share of equity, for example access to land, labour, financial resources, marketing channels or supply networks.

Equally split ownership helps to ensure that neither parties' aims and interests are quashed, by ensuring that the top management from each parent is sufficiently interested to avert problems in the framework agreement. An equal balance of equity in a framework agreement is often seen as the typical arrangement.

Owing to this balance of ownership, it can be seen that many other aspects need to balance to aid gaining success with the framework agreement. These aspects include a common language, similar background knowledge, and the sharing of long and short-term goals and objectives. Implementation then takes place using the leadership and coordination derived from the parties' similar systems.

The ownership balance is influenced by a number of different factors, namely environmental dynamics, governmental policies, organisational experience, mutual needs and bargaining position, strategic intention, investment commitment, knowledge protection, and global integration.

Ownership structure and control within a framework agreement

The balance of ownership is dependent on the relative importance of the investment to each of the individual parties. This can then be developed to imply a level of control that each party aims to hold within the framework agreement; for example, should one party wish to hold a strong control over the agreement, a majority equity share would be the probable outcome.

Increased control also allows the majority party to have a greater influence on knowledge and learning transfers. Should a larger majority be owned by a party, it will allow an enhanced control of resource application, thus increasing the control over proprietary knowledge leaks, will allow a more effective control of activities for that party's benefit and will enable a greater scope for strategy implementation.

Another aspect is the relative influence of tangible and intangible resources on the framework agreement's structure and control. Non-equity and intangible resources have been shown to have a larger influence on the control a party holds within an agreement. This can result in a minority party with the intangible resources having a stronger bargaining position than even a majority equity holder.

As can be seen, the ownership balance and structure are an integral aspect of framework agreements. Owing to the large number of elements that can influence the ownership balance and structure, the authors believe that the balance and structure of the equity applied between the parties should be derived as negotiations develop rather than stated as a figure to start.

This can be shown from the number and diversity of the determinants of the external and internal controls. The balance of ownership relative to internal controls is described through the issue of control. The party requiring the greatest internal control will probably provide the largest share of equity to gain the most influential level of control over internal determinants. External determinants will affect the relative ownership balance due to the location of the parties, whether physical or non-physical. This means that the party in the best position to control the respective external determinants should take a larger stake in equity to improve control on behalf of the framework agreement's success.

In essence, it can be seen that ownership balance and structure are about developing the best control of the numerous influences in the framework agreement. Therefore, should one party have a greater control of one area, that party should take a larger stake in the framework agreement. Then, once all influences have been researched, the 'scores' for each influence should be totalled to create an overall ownership balance and structure to be integrated within the framework agreement.

This has to be one of the most important parts during negotiations and post-negotiations owing to the definitive effect that follows. Failure in the form of starting with an incorrect balance could prove fatal.

Knowledge and learning transfers within framework agreements

Knowledge and learning transfers are necessary to make a framework agreement function. One of the main reasons parties collaborate together using the framework agreement format is to gain access to knowledge, skills or resources that cannot be created internally in a cost-efficient or time-efficient style. Nine benefits that can be gained from interpartner learning are:

1. Scrutinising partner commitment.
2. Improving knowledge flow.
3. Aligning the different cultures.
4. Building interpartner trust.
5. Integrating acquired knowledge.
6. Preventing knowledge leakage.
7. Avoidance of undue dependency on the framework agreement.
8. Establishment of reward systems.
9. Institutionalising acquired knowledge.

Interpartner learning in framework agreements

Learning is important to any business wishing to stay ahead or even alongside rivals. Without constant development and learning, companies gradually become fixed in a previous business fashion period. The main reason a business succeeds or fails is diversity, and through the learning of new skills the company will diversify, adding to its existing knowledge, and grow. To provide a constant learning pattern to constantly grow, it is necessary to create mechanisms and systems so that learning can take place.

Interpartner learning is a significant source of attainment of the knowledge necessary for expansion and gaining a competitive edge. There are numerous methods of gaining knowledge, and framework agreements are one method where quasi-internalisation (trading of knowledge) and de facto internalisation (acquiring of knowledge) can take place. It can be said that framework agreements are a method to gain cheap, fast access to new markets by borrowing a partner's core competencies, innovative skills, infrastructure and local knowledge.

The best method to learn any skill or technology is to experience it, and framework agreements can convey inter-organisational learning. Collating parties within a framework agreement with different skills, technologies, knowledge backgrounds and cultures allows unique learning opportunities. This allows the partners to the framework agreement to develop the knowledge to enhance the agreement's and their personal strategic strengths and operations.

As can be seen, this interpartner learning creates tacit and explicit knowledge transfers between the partners, where tacit knowledge is the organisationally embedded know-how that is difficult to trade, and explicit knowledge is know-how that can be easily traded through licences, franchises or the open market. Also, as stated previously, it is the tacit knowledge that creates the stronger competitive edge out of the two. Together with this, partners can bring operational knowledge, such as knowledge of technology, processes, quality control, marketing and public relations skills. They can also bring managerial knowledge, such as that of leadership, human resource management, structure, managerial efficiency, empowerment and collaborative experience. Finally, there are the financial skills that can be gained, such as skills in taxation, risk reduction, asset management, cost control and capital utilisation.

Therefore, framework agreements are a good collaborative tool because they provide an arena to combine and utilise an enhanced knowledge base to attain common goals and objectives of the partners. Without this union of knowledge the partners, if individuals, would not be able to gain the opportunity to aim for these goals and objectives.

One important point is that the creation of organisational knowledge requires dissemination of individual experiences within the network of the agreement. This means that to spread knowledge to the partners privy to a framework agreement requires a well-established system. To create a well-established system within the framework agreement, the knowledge needs to be managed in such a manner as to integrate the new knowledge with the previously owned knowledge. This will create a high knowledge survival and thus create a superior input, throughput and output of the party and the framework agreement.

Aligning the different cultures

The cultural differences that exist between partners can exist at any organisational level and are often a hindrance to learning and knowledge transfers within framework agreements. This opinion is generally correct for parties who view stability as an important factor of their structure and therefore can create barriers to learning and knowledge transfers. Open cultures, contrastingly, view change as a vehicle for learning.

As the framework agreement is drafted and negotiation takes place, awareness of cultural-interaction norms and the degree of institutionalisation needs to be expressed to aid mitigation procedures designed to resolve cultural differences. These procedures include working with partners with previous collaborative experience and, possibly, running small projects with potential partners prior to the signing of a framework agreement.

To enable cultures to blend and run in parallel together, managers need to trust each other, because suspicion of others does not create an effective learning environment. In other words, partners have to work closely to remove the problem of 'us versus them'.

Integrating acquired knowledge

Having now acknowledged the need to receive learning and knowledge transfers, a party has to be able to integrate this acquired knowledge into its existing knowledge base. A party needs to be able to have an openness to increase absorptive and integrative capacity, resulting in an increased contribution to the framework agreement along with more effective learning from the agreement and its partners.

In essence, external collaboration through a framework agreement gives access to new knowledge that cannot be generated efficiently internally, and an internal learning capacity is necessary to evaluate external knowledge.

Preventing knowledge leakage

When partners have created a framework agreement to produce new knowledge and capabilities, the risks of opportunism and knowledge leaks are very important. By creating knowledge links as part of the agreement, as well as allowing a positive flow of knowledge, the partners create the possibility that core knowledge and capabilities could unintentionally pass in the other direction.

In general it is felt this does not occur, because a framework agreement is created to combine the knowledge and capabilities of the partners. Protection of core knowledge is possible through the division of sensitive knowledge from the framework agreement, the use of contractual safeguards, an agreement by all partners to exchange only specific knowledge and capabilities, and the development of credible commitment.

Avoiding undue dependence on framework agreements

Framework agreements are used for adding to and improving a party's embedded knowledge and not to substitute for internal development. Therefore parties need to be careful when including core knowledge and capabilities to ensure that a detrimental shift in bargaining power does not take place.

A method of reducing dependence is to spread dependence to a number of partners. For example, Railtrack's Asset Protection Department utilised three engineering consultants, namely Jarvis Rail, Corus Rail Consultancy and Atkins Rail. All three consultants were privy to a framework agreement between themselves and Railtrack. Railtrack then assigned work packages subject to the present confidence in the consultant, to create a structured environment where a consultant was always available for work, creating a constant need to develop from the others.

Establishment of reward systems

To integrate new knowledge into the framework agreement, reward systems are necessary to provide an incentive to integrate new knowledge. Lei *et al.* (1992) have defined two reward systems that could be used within framework agreements, namely hierarchy-based and performance-based systems. Hierarchy-based systems develop strong links between the partners and their people by defining formal organisational boundaries. Performance-based systems put a premium on quantitative measures of performance. The incorporation of these can create a suitable incentive system to increase the collection of learning and knowledge transfers, and to strengthen the party's specific knowledge.

Institutionalising acquired knowledge

Within the framework agreement and the partners' systems, it is necessary to transform the new knowledge and capabilities learnt into the party's specific knowledge. Therefore, the institutionalising of acquired knowledge is the process of changing information into knowledge and individual experiences into organisational lessons.

To attain this ideal it is necessary to include, within the framework agreement and the partners' structure, venues for knowledge and capabilities to be collected and then passed to the relevant area of the business. This creates a feeder system that is transparent and functional.

Summary of knowledge and learning transfers within framework agreements

Knowledge and learning transfers are the fuel that drives the engine, figuratively speaking. Without knowledge and learning transfers a framework agreement could not function and instead the partners would be a group of companies with plenty of ideas, but no form to structure them. The use of knowledge and learning transfers allows parties to temporarily or permanently acquire knowledge in its various forms, in a more efficient and effective manner than internal development.

Further analysis shows that knowledge and learning transfers help to create the structure of a framework agreement. Without using knowledge and learning transfers it is not possible to combine any of the enabling elements of the partners such as goal compatibility, cultural traits and organisational traits. It is through the knowledge and learning transfers that, once those enabling elements of the partners have been combined in the framework agreement, collective progress can be made.

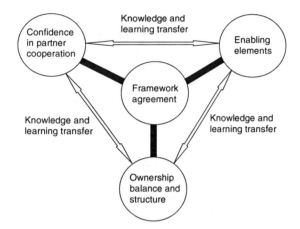

Figure 11.2. A model for framework agreements

Another positive that can be drawn from the knowledge and learning transfers is the building of confidence in partner cooperation. The openness between parties necessary to develop knowledge and learning transfer links shows that a certain element of trust must exist through communication and the interpartner adaptation that must occur as a result of the passing of partner-specific knowledge to another. The control of knowledge leakage will also help to develop confidence in partner cooperation, because the partners will be able to govern what knowledge and learning are transferred to partners within the framework agreement.

It is also through the knowledge and learning transfers that ownership balance and structure can be defined. It is the holder of knowledge that can have a large impact on the balance of power within the agreement. The parties with large banks of internal knowledge hold a stronger bargaining power than those without. A party requiring knowledge gain will look for not only a knowledgeable party to learn from, but a party that can transfer the knowledge to the first party effectively and efficiently, reducing the transaction period and allowing more knowledge and learning transfers within a shorter period of time.

A model for framework agreements

Now that we have determined the various influential constituents of a framework agreement, it is time to combine them into a model (Figure 11.2). The influential constituents are the enabling elements (as shown in Figure 11.1), confidence in partner cooperation, and ownership balance and structure. It is felt that all these constituents have an equal influence on the operational efficiency and effectiveness of a framework agreement.

All three constituents are interrelated and exist in a multi-project environment or long-term relationship. The relationship between these three exists because the quantity and quality of the enabling elements affect the ownership balance and structure, and could raise or lower confidence in partner cooperation. Should the confidence in partner cooperation change, a party may deem it necessary to alter its strategic approach through its enabling-element provision, and this could have a secondary effect regarding the ownership balance and structure. Likewise, adjustments to the ownership balance and structure have subsidiary effects on the other two constituents.

It is the knowledge and learning transfers that provide the vehicle for these relationships. Without the passing of knowledge around the model, partners to the agreement would not be able to judge any of the three constituents so as to provide conclusive responses relative to the actions of the other partners. This would mean the framework agreement, if possible to create without knowledge and learning transfer, could make no progress forwards (or backwards) upon commencement.

Enabling elements

It is through the eight elements shown in Figure 11.1 that the complexities and reliance on the framework agreement are created. The enabling elements also aid the definition of any compatibilities and complementary resources that exist.

Through the use of these enabling elements, historical experiences in the form of industrial and business background, future proposals, and marketing and distribution networks that the parties have already developed are taken into account. This results in an opinion of market power and bargaining position already held as a result of previous transactions. Therefore, this model includes an infinite timescale, allowing it to be used at any period of the life cycle of any project(s) or programme(s), and can include any previous or future influences, no matter how remote.

Most importantly, for any framework agreement, the enabling elements identify the goals and objectives of the contractual arrangement. If the goals of the partners are not collectively compatible, it would be more efficient to address these goals individually rather than through a framework agreement. The model also provides assessment capability and the use of complementary resources to achieve the compatible goals of the framework agreement, thus creating a fully dimensional supply and demand scenario, where the compatible goals are the demand which is supplied by the complementary resources, commitment, and a capability to supply.

From the above, it will be these that help to construct confidence in partner cooperation and aid the development of ownership balance and structure. These are the two other constituents and thus there is a need for equilibrium.

Confidence in partner cooperation is linked to the fluctuation of trust and control. Trust can be built from any of the enabling elements and is an opinion of your partners' collaborative drive in this instance. The most influential element on trust levels is commitment, for obvious reasons. A party's control is derived from those elements that make possible empowerment and lay down the boundaries and parameters. From the organisational, strategic and cultural traits of the framework agreement, a party will have a control level over partners that will help to determine its confidence in partner cooperation.

Dependent on the structure of the enabling elements, this will determine what ownership balance and structure exists. During the negotiation period, each party will have hunted for a potential partner that could help in the construction of a framework agreement with an ownership balance and structure in mind. It can be seen that the elements contain equitable/inequitable and tangible/intangible resources, goals, capability, knowledge and so on. Through this model, any possible and feasible influence on ownership balance and structure can exist, creating an open, highly adaptive model to be utilised in any environment that a framework agreement could experience.

Confidence in partner cooperation

The second constituent of the model to be discussed is confidence in partner cooperation. This is the differential driveshaft of the framework agreement vehicle. Should confidence be high, the framework agreement will run smoothly, and attain all goals and objectives of the partners and the agreement. If confidence is low, goals may not be achieved, commitment to the agreement could reduce and peer pressure increase.

The utilisation of control mechanisms aims to organise the enabling elements and the ownership balance and structure so that the framework agreement can be run in a productive manner for all parties and thus increase confidence. The control mechanisms need to be adaptable to achieve the necessary formal and social-control successes, because the influential enabling elements are generally dynamic and change over time.

Trust, like control, has an influence on the confidence in partner cooperation and, in turn, the enabling elements and the ownership balance and structure. Should trust increase, confidence in the framework agreement will increase and the party concerned will be more committed to the agreement. Increasing commitment means that the party will tune its organisation, strategic objectives, cultural adaptations and resource complement more towards goals and objectives compatible with the other partner and the agreement.

Trust levels also affect the ownership balance and structure. A high level of trust will improve partner adaptability and the willingness to modify the ownership balance and structure for the benefit of the agreement and partners' needs. Without high confidence in partner cooperation, parties will be suspicious of any suggested alterations to the ownership balance and structure. This suspicious attitude could result in a breakdown in trust and confidence overall, thereby terminating the ownership balance and structure, together with the framework agreement.

Ownership balance and structure

Finally, to complete the picture, the influences of ownership balance and structure on the other constituents will be discussed. As with the other two constituents, the ownership balance and structure are integral to the success of a framework agreement.

Ownership balance and structure have similar basic building blocks to the other constituents, but have another, different method of finalising those blocks to produce a structure. Ownership balance and structure provide a delivery system to allow the framework agreement to function.

The ownership balance and structure provide a statement of proportional delivery of the enabling elements. What this means is that the ownership balance and structure can usually be related to the level of commitment, the provision of resources, the organisational structure of the agreement, the financial contributions and so on. Therefore, dependent on the negotiation procedure, the ownership balance and structure can govern what each partner must contribute towards the framework agreement.

The ownership balance and structure affect the confidence in partner cooperation by creating a standard measurement for the framework agreement. The ownership balance and structure show the expected work level, provision level and anticipated grants relevant to the framework agreement. Should each partner achieve its necessary investment levels, the confidence in partner cooperation will increase because all partners are apparently committed to the framework agreement. Therefore, there is a clear relationship between the ownership balance and structure, and the confidence in partner cooperation.

Now that we have discussed the three constituents of a framework agreement and the relationships between them, the role of knowledge and learning transfers will be included in the model.

Knowledge and learning transfers

The knowledge and learning transfers provide the medium for interaction between the constituent elements of the model. Without knowledge and learning transfers there is no other method of creating an interaction interface between the constituents, and without this interaction the framework agreement cannot function as intended. Instead the framework agreement would be a contractual system for collecting attributes, rather than a good-attaining, integration body to be utilised by collaborative partners.

The knowledge and learning transfers are not only a medium for passing information, but are a feedback channel for the partners involved. For example, should the development of enabling elements take place, and thus confidence increase, the knowledge and learning transfers allow the partner(s) of the framework agreement to inform the enabling-element-developing partner that they have made a good, progressive move for the agreement and would like a continuation. The knowledge and learning transfers also allow negative directives to pass. Say, for example, a partner wants to reduce its equity balance within the framework agreement; this will be communicated to the other partners, reducing confidence and redistributing the provision of the enabling elements.

The knowledge and learning transfers also allow progress of the framework agreement as a whole. The knowledge and learning transfers allow the new skills and attributes acquired by the partners to become a party's new specific knowledge. The party can now, through the knowledge and learning transfers, again acquire new skills and attributes to be added to the ever-expanding party-specific knowledge.

Alliance and framework agreements in the public sector

The traditional approaches to executing construction projects are becoming increasingly unsuitable for meeting the demands of most projects, as a result of increasing technical complexity, value and risk. In order to meet or satisfy such demands, a high level of coordination and flexibility is expected, to minimise cost and time inflation on the one hand and increase quality and safety standards on the other. There are certain factors that must be considered if a public sector agency/outfit (hereafter referred to as the client) decides to partner. Government policy regarding procurement is that it should be based on best value for money and that all public entities should seek to secure continuous improvement in value for money. Best value for money means taking into account the optimum combination of whole-life cost and quality necessary to meet a public entity's requirement. Thus all procurement decisions should be based on robust assessments of all the options in each set of circumstances throughout the life of a contract through effective contract monitoring and control.

In most cases works are procured by means of open competitive procedures. Two-stage tendering, selective tendering and single-source procurement are also possible under strict conditions that necessitate their choice. In order for framework agreements to be practicable under these conditions, it must be seen that competition and accountability are not sacrificed. Hence such projects must be competitive at the outset in the selection of the alliance partner. There should also be clear definition of the contractual responsibilities of both parties and specified and measurable milestones for improved performance of the contract. Since the public sector would want to give an equitable and fair opportunity to all contractors, the relationship can only be for a specified period of time.

To start with, the client must determine its reasons for using a framework agreement. This is the first stage in the process and must therefore clearly and unambiguously define its needs, and answer the question of whether a framework agreement will fulfil them. This stage should also involve an internal assessment of the client organisation and of how an alliance might benefit existing strategies. For the public sector, its business drivers could include minimising engineering and construction cost, as well as the maintenance of the facility (whole-life costing) and reducing the frequency of projects going to litigation (and resulting in delays and cost inflations).

If the client expects improved performance through project execution (value for money), then the potential benefits to be realised should also be substantial, since the success of the capital investment programme will determine the government's overall success for a long time in the future. This means that the client should be prepared to invest in and allocate skilled personnel to meet this expectation on large projects that can bring about substantial benefits or returns.

The decision to form an alliance should also be guided by the uncertainty and risk factors affecting the project. If projects are complex and the risks (in terms of technology, location, time, quality and cost) are considerable, then an alliance should be considered as a powerful mitigating tool or concept to deliver the project. As opposed to the traditional form of contract, an alliance should provide a better arrangement for analysing risks and dealing with their consequences. It does not merely transfer risk to another party but rather allocates it to the party that is best suited to manage it.

The selection procedures used by a public sector authority should follow the procurement guidelines set out, as they are governed by rules that must be strictly enforced. Public procurement policy and procedures are meant to provide the country with economy, open competition, transparency and accountability, as well as a balance with any national policy of enhanced development of local industries.

The decision to enter into an alliance should be based on an assessment of the work to verify that it will provide better value for money and also ensure fair and transparent allocation of risk. The client must be convinced that an alliance with a framework agreement will achieve continuous improvement and reduce confrontation. The selection of contractors by the public sector has traditionally been based on competitive tender to demonstrate probity and value for money. The competition has been defined by the lowest tender price and this brings with it a lot of problems. In most cases the lowest prices contain no margin of profit for the contractor, whose commercial response is then to try and claw back the margin which was not in the tender through variations, claims and 'squeezing' of suppliers and subcontractors. The formation of the alliance should take into account that reasonable margins must be allowed and aligned objectives agreed when the framework agreement is signed. Measurable targets are set at this stage, and incentives for savings and improvements also agreed. A dispute resolution structure and procedure must be clearly set out, so that all issues or problems can be dealt with as soon as they crop up or avoided if they are foreseen. Regular reviews of completed sections of work or sub-projects must be carried out to compare achieved results, particularly with respect to time and cost, against targets set.

Summary

A framework agreement is a support for any number of transactions, especially in a multi-project environment or long-term relationship where a large number of transactions take place. In a framework agreement, three constituents need to be fully and equally satisfied to ensure that a strong framework agreement results. Should one of the constituents be neglected, this will have the effect of creating an unstable framework agreement, and should a constituent be missing, either the framework agreement will never be formed or it will collapse if already in existence.

In the framework agreement model, the knowledge and learning transfers ensure that a relationship exists between all three constituents, providing a system to regulate their influence.

Bibliography

Arino, A. Veracity and commitment: cooperative behaviour in first- time collaborative ventures. In: Beamish, P. W. and Killing, J. P. (eds), *Cooperative Strategies*, vol. 2. *European Perspectives*, pp. 215–241. New Lexington Press, Indianapolis, IN, 1997.

Blodgett, L. L. Partner contributions as predictors of equity share in international joint ventures. *Journal of International Business Studies*, **22**(1) (1991), 63–78.

Buckley, P. J. and Casson, M. A theory of cooperation in international business. In: Contractor, F. J. and Lorange, P. (eds), *Cooperative Strategies in International Business*, pp. 31–53. Lexington Books, Indianapolis, IN, 1988.

Creed, W. E. D. and Miles, R. E. Trust in organisations: a conceptual framework linking organisational forms, managerial philosophies, and the opportunity costs of control. In: Kramer, R. M. and Tyler, T. R. (eds), *Trust in Organisations: Frontiers of Theory and Research*, pp. 16–38. Sage, London, 1996.

Das, T. K. and Teng, B.-S. Between trust and control: developing confidence in partner cooperation in alliances. *Academy of Management Review*, **23**(3) (1998), 491–512.

Dasgupta, P. Trust as a commodity. In: Gambetta, D. (ed.), *Trust: Making and Breaking Cooperative Relations*, pp. 49–72. Basil Blackwell, Oxford, 1988.

Doz, Y. L. The evolution of cooperation in strategic alliances: initial conditions or learning processes? *Strategic Management Journal*, **17**(Special Summer Issue) (1996), 55–83.

Doz, Y. L. and Hamed, G. *Alliance Advantage*. Harvard Business School Press, Boston, MA, 1998.

Geringer, J. M. *Joint Venture Partner Selection*. Quorum Books, London, 1988.

Geringer, J. M. Strategic determinants of partner selection criteria in international joint ventures. *Journal of International Business Studies*, first quarter (1991), 41–62.

Goold, M. and Quinn, J. J. The paradox of strategic controls. *Strategic Management Journal*, **11** (1990), 43–57.

Gulati, R. Does familiarity breed trust? The implication of repeated ties for contractual choice in alliances. *Academy of Management Journal*, **38** (1995), 85–112.

Hamel, G., Doz, Y. L. and Prahalad, C. K. Collaborate with your competitors – and win. *Harvard Business Review*, **67** (1989), 135–139.

Kanter, R. M. Collaborative advantage: the art of alliances. *Harvard Business Review*, **72**(4) (1994), 96–108.

Lei, D., Slocum, J. W. and Pitts, R. A. Building cooperative advantages: managing strategic alliances to promote organisational learning. *Journal of World Business*, **32**(3) (1992), 203–223.

Locke, E. A and Latham, G. P. *Goal Setting: a Motivational Technique That Works*. Prentice-Hall, Englewood Cliffs, NJ, 1984.

Park, S. H. and Russo, M. V. When competition eclipses cooperation: an event history analysis of joint venture failure. *Management Science*, **42** (1996), 875–890.

Sitkin, S. B. and Stichel, D. The road to hell: the dynamics of distrust in an era of quality. In: Kramer, R. M. and Tyler, T. R. (eds), *Trust in Organisations: Frontiers of Theory and Research*, pp. 196–215. Sage, London, 1996.

Zeira, Y. and Shenkar, O. Interactive and specific parent characteristics: implications for management and human resources in international joint ventures. *Management International Review*, **30**(special issue) (1990), 7–22.

CHAPTER TWELVE

Innovative procurement methods

S. Male

Introduction

This chapter takes as its starting point the recent report *Accelerating Change* (Strategic Forum for Construction, 2002), which has set the following strategic targets:

- 20% of all construction projects by value should be delivered using integrated teams and supply chains by the end of 2004. That figure should rise to 50% of projects by value by the end of 2007.
- 20% of client investment in projects by value should adopt the principles of the Clients' Charter by the end of 2004, with this figure rising to 50% by value by the end of 2007.

These are challenging targets. *Modernising Construction* (National Audit Office, 2001) highlights the fact that the Office of Government and Commerce (OGC) advised central government departments that, as from 1 June 2000, they were to secure construction projects using three primary procurement routes: public–private partnerships, especially the Private Finance Initiative (PFI); design and build; and prime contracting. The OGC has also advised government clients to adopt these three routes for refurbishment and maintenance activity as from 1 June 2002. The OGC indicates that traditional non-integrated procurement options should only be used if they are able to demonstrate best value for money, with the expectation that they would seldom be used in practice.

The chapter explores the recently launched initiative of the Clients' Charter and the meaning of integrated teams and supply chains, and compares different procurement routes against the concept of the project value chain, using the traditional procurement route as the benchmark. Client types and client value systems are introduced, including exploring the relationship of maintaining a slender 'value thread' intact throughout the project delivery process and its implications for different procurement strategies.

The chapter raises fundamental questions about what is 'value for money' and for whom in construction procurement.

The Clients' Charter

The Latham Report (Latham, 1994) argues that clients are at the core of the project delivery process and are the driving force behind any agenda for change. The Deputy Prime Minister, John Prescott, in July 2000, challenged clients in construction to draw up a charter setting out their minimum standards expected from procuring projects currently and their aspirations in this area for the future. The charter is also one for client self-improvement. Charter clients are drawn from the public and private sectors, and can be large, small, regular or irregular procurers of construction. A management board manages the charter process, with membership drawn from the demand and supply sides of the industry. The board is serviced by and located within the structure of the Confederation of Construction Clients (CCC). Charter clients pay an annual registration fee to the CCC. A measurement toolkit is used to collect a consistent set of data to underpin client self-measurement and benchmarking.

Client members of the CCC must commit to the Clients' Charter, which requires that they are obligated to improve their performance in four key areas:

- *Leadership and focus on the client.* This component of the charter requires clients to commit to providing leadership to improve the procurement process, to identify and manage risk, and to encourage supply side service providers to innovate to meet their requirements. Leadership responsibilities for clients also require that they set clearly defined objectives, which ideally should be quantifiable, and with realistic targets. There is a commitment to engendering trust, team working and a non-adversarial approach throughout the supply chain, with suppliers treated fairly, including using appropriate payment procedures. Partnering is advocated as the favoured approach for delivering construction projects.
- *Product team integration.* This aspect of the charter commits clients to promoting sustainability in construction, long-term relationships with the supply chain such that all relevant parties can be involved in the design process, and maximising the benefits of standardisation and off-site fabrication.
- *Quality.* This aspects of the charter commits clients to search for quality-based solutions that take account of and respect the

surroundings, maximise functionality and optimise whole-life cost whilst minimising defects through process and product improvements.

- *People*. Engendering respect for people throughout construction activity and having a commitment for clients to train their own staff.

In addition to the four areas for cultural improvement in the industry, charter clients are also committed to adopting national key performance indicators for benchmarking, namely:

- client satisfaction with products and service
- defects
- time and time predictability
- cost and cost predictability
- safety
- profitability of the supply chain
- productivity.

Charter clients are aiming to meet or better the targets set in *Rethinking Construction* (Department of the Environment, Transport and the Regions, 1998) and provide a focus for incentivisation around appropriate parameters, as discussed earlier in the book.

Whilst the CCC does not commit its membership to using a particular procurement route, it recommends partnering, which is not in itself a procurement route. It is an approach to supply chain integration, the subject of the next section.

Integrated teams and supply chains

Integrated teams and supply chains have been the focus of a number of recent major reports. The *Accelerating Change* report views these as at the heart of delivering client value. Its authors expect clients to appoint established, integrated teams and supply chains that are used to working together. This creates the capability of moving from project to project to engender a culture of learning and continuous improvement using performance measurement.

Integrated teams involve bringing the right skills together at the right time in the project process, regardless of where they might be located in a supply chain. They will involve clients and those responsible for the delivery process, including manufacturers and other types of supplier. Integrated teams are founded on trust and mutual respect, have a strong client focus and operate through the equitable sharing of risks and rewards using appropriate incentive mechanisms. Processes supporting team functioning that rely on information technology to assist integration are seen as an important underpinning for their successful operation. The involvement of manufacturers, suppliers and specialists is seen as a critical issue for integrated teams. They will be able to contribute to developing solutions that improve on-site methods of working, increase standardisation, permit increased levels of off-site pre-assembly and prefabrication, reduce risks, especially in the area of health and safety, and improve reliability and quality. They are also seen as bringing to the process their research and development expertise, to the benefit of clients and the industry in general.

'Building down barriers' (BDB) is a joint approach to innovative procurement combining client-led expertise from Defence Estates (DE), the Ministry of Defence (MoD) and the then Department of the Environment, Transport and the Regions to establish mechanisms for supply chain integration in construction. The initiative was launched in 1997, and two pilot studies using the new procurement method were completed in late 2000. The BDB approach focuses on providing a structured process and collaborative models of leadership to integrate pre-assembled supply chains under a prime contractor. Prime contracting will be discussed further below; however, the generic approach to BDB is that single-project relationships are replaced by multi-project relationships based on trust and cooperation. Designers and constructors are brought together under single-point responsibility. The intention is that supply chains remain together over time. Holti *et al.* (2000) identify three types of leadership essential for supply chain integration:

- Supply chain leadership giving single-point responsibility to the client and providing overall leadership for achieving value for money. This is the role of the prime contractor.
- Design leadership, focusing on extracting, through dialogue with the client and end-user, project values, functional requirements and other design parameters; and on developing a design strategy which is consistent with the foregoing and ensuring that design development remains consistent with project values. An essential skill here is design management to ensure separate design activities are programmed, coordinated and integrated.
- Construction/delivery leadership, focusing on developing a construction strategy that is consistent with project values and on coordinating inputs from manufacturing, construction and facilities management expertise to deliver the design within a target whole-life cost. An essential skill here is that of construction management, namely planning, monitoring and resource acquisition and implementation to ensure work packages, trades and organisational interfaces work effectively.

The next section will review issues surrounding clients and their value systems and their implications for integrated procurement methods.

Client types, client value systems and procurement

Large, regularly procuring clients are increasingly pursuing innovative approaches to the way in which their projects are planned, procured, designed and delivered. Procurement is seen as an adjunct to facilitate delivery of their business strategies. They are looking to work closely with supply chains to maximise value and achieve continuous improvement in construction performance. Clients to construction projects are much more diverse than the segment under discussion here. A number of distinguishing characteristics can be applied to clients, providing insights into their different value systems.

Mention has already been made of them separating into those in the public and private sectors. Public and private sector clients have different value drivers, not least public accountability in the case of the former and the fact that they also have to take account of legislative influences on procurement at European Community level. Private sector clients are much more heterogeneous. Markets, the Stock Exchange, shareholder value and ownership considerations due to plc, private limited company or perhaps family business status can also act as different 'value drivers' in the private sector.

Equally, clients also differ in their level of knowledge of the construction industry and its operations. Knowledgeable clients normally adopt a very structured approach in dealing with the industry and project delivery, usually described in a project manual covering procedures or guidelines. Knowledgeable clients will treat the construction supply chain and its members as 'technical experts' to deliver a project or projects to meet their business and/or social need. They will place considerable demands on members of the construction industry and expect it to respond accordingly. Internal or external project managers will act as the interface with the construction industry. Knowledgeable clients will tend to be innovative with procurement methods and are generally the volume procurers of construction services. However, client knowledge of the industry and its processes can be considered as a continuum. For example, there are those clients that have a deep, diffused knowledge and understanding of the industry across the entire organisation. There are those clients who have an extensive knowledge of the industry but it is located and remains within a division, department or unit that acts as a boundary manager and gatekeeper between the internal organisation and the external construction industry. At the other end of the continuum are less knowledgeable clients who approach the industry infrequently or on a one-off basis. They are irregular procurers and will have limited or minimal in-house expertise and knowledge of the operations of the industry and its complexities. Evidence suggests that this type of client may be directed into a traditional procurement path, depending on their initial contact with the industry. Depending on their business networks, they may also consult clients that are more knowledgeable.

Clients as customers and users of construction can also be classified as large or small owner/occupier clients. Large owner/occupier clients require physical assets to support their ongoing business or socially driven needs and agendas. There are also small owner/ occupier clients, who tend to be reactive and approach the industry because their existing facilities are inadequate in some way. Developers approach the industry and construction to make a profit from the sale, rent or leasing of facilities. They trade in physical assets or investment in them to generate profit.

Clients can also be characterised by the demands placed on the industry in terms of the type of construction, the volume of activity, its frequency and its regularity. This can be linked to the extent to which standardisation exists from project to project in terms of parts, processes and design. Unique construction occurs when it is distinctive in terms of technical content, the level of innovation required or the extent to which it is a leading-edge project which pushes the boundaries and envelope of the industry's skills and knowledge. Unique construction has limited scope, if any, for efficiencies in processes or standardisation and repetition. Off-the-peg construction has similarities to unique construction but the possibility exists for standardisation through repeat designs across a few buildings, customised construction is a more apt description, since designers may take previous designs, perhaps undertaken for other clients, and adapt them to the situation at hand. Much of the industry workload is reflected in this type of activity. Process construction occurs where a client has repeat demands for projects permitting a high degree of standardisation due to volume procurement. Efficiencies can be made from standardisation of design, components and processes. Cross-project learning is also possible. There are similarities to the manufacturing sector. Process construction could also include instances where there is a relative balance between 'new build', maintenance, refurbishment and retrofit of existing assets. Clients may be volume procurers in terms of maintenance and refurbishment of existing assets but be *ad hoc* procurers of new-build work. Finally, portfolio construction occurs where clients are large and regular procurers with substantial investment programmes requiring different types of construction activity. A diverse range of needs will exist for clients involved in this type of activity. Portfolios will comprise different technical requirements, and different degrees to which construction activity is of a unique, customised or process nature. Regular, ongoing investment programmes permit long-term relationships to be developed with suppliers.

A client typology is presented in Table 12.1.

In summary, clients to construction are heterogeneous. Large, regular procurers are driving change deep into the industry. Organised groups of clients are also driving change into the industry. However, they do not represent the full diversity of clients procuring construction. Whilst generic influences on client value systems have been identified in Table 12.1, each client has its own distinctive value system, derived from its structure, cultural web, ownership characteristics, and strategic and operational management processes. The value system and value drivers will be influenced by the client's sector of operation, its organisational structure and functioning, and the manner in which it approaches and engages with the industry. Additionally, clients to construction may procure a single project, or they may procure a number of projects over a prolonged period which are grouped into a project programme as discussed in Chapter 13. Again, this will influence and affect the client's value system and value drivers.

The next section explores the concept of the project value chain as a mechanism to bring together project development and delivery, procurement strategies, and client value systems and value drivers.

The project value chain

Organisations should be considered as a series of internal and external activities that comprise a 'value chain' giving a competitive edge in the market place. Value chain activities provide the basic organisational infrastructure for creating and delivering value. The concept of the value chain has now been extended into the discipline of managing by projects. The project value chain is used to view projects as a series of value-adding activities that have their origins in and emerge from the client's statement of business need. To deliver projects successfully, these activities have to remain in alignment into the operational and use phase to ensure the product provides fitness-for-purpose as an outcome of project processes. In this sense, project value chain activities comprise part of the broader value chain activities of an organisation. Value management and

Table 12.1. *Client impacts on the construction industry*[a]

	Client type									
	Private sector clients						Public sector clients			
	Knowledgeable – regular procurers			Less knowledgeable – infrequent procurers			Knowledgeable – regular procurers		Less knowledgeable – infrequent procurers	
Response to the industry	Consumer clients: large owner/ occupiers	Consumer clients: small owner/ occupier	Speculative developers	Consumer clients: large owner/ occupier	Consumer clients: small owner/ occupier	Speculative developers	Consumer clients: large owner/ occupier	Consumer clients: small owner/ occupier	Consumer clients: large owner/ occupier	Consumer clients: small owner/ occupier
Unique				✓					NA	NA
Customised	✓	✓	✓	✓				✓	NA	NA
Process	✓	✓			✓		✓	✓	NA	NA
Portfolio	✓		✓				✓		NA	NA

[a] A tick denotes that this is a probable occurrence. NA indicates no occurrence and a blank indicates a possible but unlikely occurrence

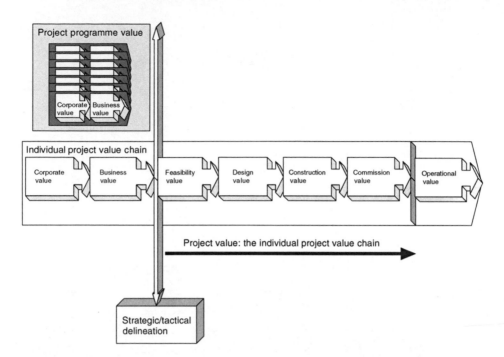

Figure 12.1. The project value chain, adapted from Standing (2000)

value engineering have the capability of aligning or realigning the project value chain. They are also mechanisms to keep the 'value thread' intact through strategic consideration of procurement as part of the value management process. Procurement and contract strategies are important strategic decisions that may maintain the alignment of the project value chain or create barriers or discontinuities within it. The project value chain is set out schematically in Figure 12.1.

The decision to build/construct is an important strategic value point for the client within the project value chain. The project is effectively outsourced to the construction industry. Whilst Figure 12.1 presents the decision to construct as a discrete event, Woodhead, with his work on the 'decision to build', and Graham, in the area of the private financing of water infrastructure projects, indicate that it is much fuzzier in practice. This chapter views the decision to construct as a business commitment that a project requiring the skills of the construction industry is the right solution and that capital funding is being made available for further investigation. Depending on the client type and its structure, the early, strategic phase of the project value chain may involve considering more than one project. Projects may be competing with each other for investment. This requires them to be managed as a holistic framework at the project programme level, with 'value' being considered from multiple- or single-project perspectives and often simultaneously.

Transition points occur in the project value chain, such as the 'decision to construct/build' and at the 'handover, commission and operational user interface', for example. Discontinuities could occur at any of the transition points because there are changes in the 'value structure' of the project due to the influence of the organisations involved, changes in organisational involvement or a different focus being applied to the project. Equally, as a project goes through each of the phases of its development, other value transition points occur, for example at feasibility stage, design stage and construction stage. The manner in which these value transition points relate to each other and are managed is impacted by the procurement strategy.

The next section will explore the consequences of the project value chain for procurement strategies.

Procurement and the project value chain

The previous sections have highlighted moves towards more integrated approaches to procurement, with the PFI, prime contracting, and design and build being pushed extensively by central government clients. As mentioned earlier, the CCC advocates partnering as the favoured approach to projects. The choice of procurement route is intrinsically linked strategically with client value drivers. Procurement and contract strategies are not tactical choices within projects. They involve decisions on the managerial and legal frameworks set up for risk allocation, the delivery of functionality in the design and construction stages, and the relationship between time, cost and quality. Examples of these value criteria are set out in Figure 12.2.

A more multifaceted view of project value drivers is given in Figure 12.3. It highlights that delivering value is at the centre of a whole series of decisions and contextual influences. These have to be fully understood when making choices about procurement.

An earlier chapter has described in detail the characteristics of a range of different procurement routes available to clients. Figure 12.4 schematically presents the major procurement systems against the project value chain. This section will explore each of the systems, noting issues that are raised through comparison with the project value chain and the value drivers framework noted above. The thick diagonal lines in the figure denote major value transition points.

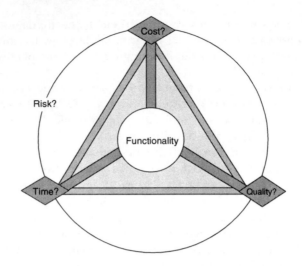

Figure 12.2. Examples of project value drivers, adapted from Standing (2000)

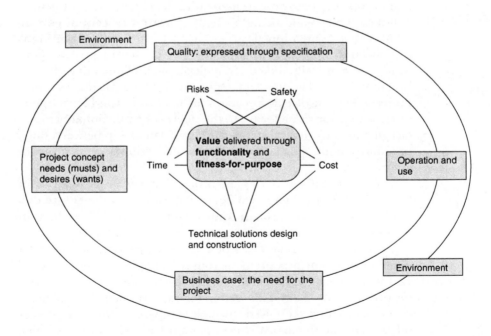

Figure 12.3. A value driver framework for procurement

The argument put forward here is that providing the project value chain remains in alignment throughout, such that each stage builds on the others with a 'value thread' remaining intact and consistent from stage to stage, value will be created and delivered for the client. In general, procurement systems at the top of the diagram provide more opportunity to maintain the integrity of the project value chain since an increased number of discrete activities come under one umbrella organisation for single-point delivery. There is one proviso: they must be designed and delivered with that intent in mind.

The traditional procurement route at the bottom of Figure 12.4, used here as a benchmark, is potentially the most disruptive to the project value chain. Where single-stage competitive tendering is adopted for traditional procurement, the project value chain is disrupted completely at the transition point between design value and construction value. Single-stage tendering under the traditional procurement system necessitates that tender documentation is able to fully capture the client value system embedded in the design up to tender stage. Where two-stage tendering is adopted, the capability exists of bringing construction expertise into the project much earlier and improving integration. It also offers the capability of introducing some fast tracking.

There is an interaction between the procurement method, the choice of tendering strategy and the potential impact on the integrity of the project value chain. In addition, the traditional approach is essentially sequential. This provides it with one of its strengths, namely that the design can be fully worked up and tested, with the client value system embedded into the design through a dialogue with the designers. The tendering process provides it with one of its weaknesses, since it does not permit constructor and supply chain involvement in design. Constructors build to contract and a limited number of multiple constructors are in competition with each other, assuming selective tendering has been adopted. It is argued strongly that the traditional route is adversarial, is the cause of contractual conflict and, as noted above, should no longer be adopted for central government procurement of construction unless a good-value case can be made. Latham (1994) noted, however, that the briefing process leaves much to be desired, which is emphatically a client issue and can lead to many problems later in the project, if not addressed adequately and sufficiently early on in the process. Equally, tender documentation can often leave much to be desired, with designs often incomplete before tender. The industry has progressively worked out standard contracts for the traditional system that have also permitted risk allocation to be clearly identified

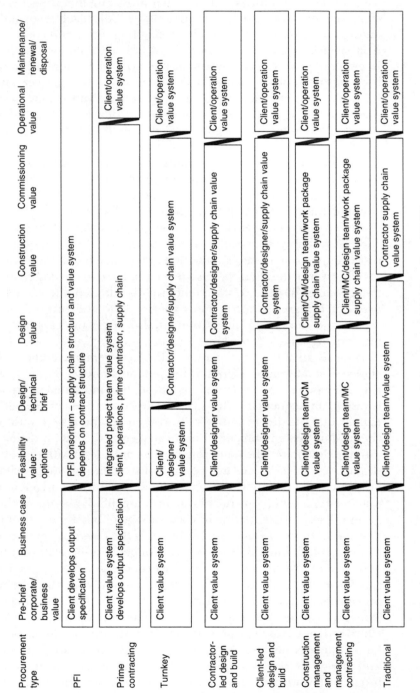

Figure 12.4. Procurement strategies and the project value chain, adapted from Standing (2000)

for all parties. Design is only paid for once. Historically, until a Code of Selective Tendering was brought in, the traditional system was seen as adding expensive tender costs into the industry, although a number of firms are still producing bids even under selective tendering. Provided the system is adhered to and designs are fully worked up and tested through cost planning, it does provide the client with cost certainty. The traditional route also provides opportunities for quality designs to be produced and built but has also been criticised for its relative lack of time certainty. An opportunity exists to provide some integration within the traditional procurement route by using value management and value engineering at various stages in the process, and bespoke partnering arrangements can also be overlaid onto standard forms of contract, which include bringing in early supply chain involvement. This is not, however, without some difficulties related to contractual appointment of specialist subcontractors and suppliers. The traditional route assumes sequential procurement of specialist supply chain members. The traditional route is recognised widely internationally.

Management contracting (MC) and construction management (CM) are attempts at increasing the level of integration within the project delivery process. The allocation of risk differs between the two, with the client picking up work package risks with CM whereas they are allocated to the management contractor under MC. Both have the capability of increasing team-working on a project at important interfaces. They offer enhancements to the traditional approach in terms of integrated team-working, with additional constructor management and buildability knowledge brought into the design team consultant domain. Both have the capability of fast-tracking the delivery process. The client value system has an opportunity to be embedded into the design process in much the same way as traditional procurement. Again, design is only paid for once, and the protagonists of CM and MC argue that these methods permit the designers to design and take away the design management and coordination responsibilities from them. CM and MC are often criticised for their lack of cost certainty for the client. The management forms of procurement permit increased involvement of construction knowledge earlier in the process, but essentially they are profession-led routes, with a consequent increase in the number of interfaces to be managed, with the potential for project value chain disruption to occur. They are placed towards the bottom of the project value chain for integration but are considered better than the traditional system. They can be overlaid with value management, value engineering and partnering systems to improve integration further.

There are numerous variants on design and build (D & B), including those offered by negotiation, and by single-stage and two-stage tendering. The client and design team can work up a substantial amount of the design, for example up to sketch design, and then tender to a D & B contractor to develop the detailed design. The client's design team can be novated to the D & B contractor, or the client can develop a brief and then go out to a D & B contractor to undertake full design development. Competitive D & B can be disruptive to the project value chain for a number of reasons. First, tender documents have to be well written to provide insights into the client value system. Second, D & B contractors during tender do not have direct access to client thinking about their requirements and this cannot be fully encapsulated within design development work until after tender award. Third, members of the contractor's supply chain may be prevented from having early access to the client value system, depending on when competitive tendering takes place. They do, however, have access to the D & B contractor's designers. Depending on the mechanism for implementing D & B, it is highly likely that more than one design will be developed as part of the tendering process, with the associated tendering and design costs for unsuccessful bidders having to be recouped from somewhere. The choice of system adopted, including negotiated, single-stage or two-stage tendering, permits different levels of the client value system to be embedded into the project prior to and post tender. However, as in previous examples, wherever a tendering requirement exists, the standard of documentation produced will determine the extent to which client value requirements are easily understood and encapsulated into further project delivery stages. D & B procurement offloads risk onto the D & B contractor through single-point responsibility for coordinating design and construction. Again, value management, value engineering and partnering can be overlaid onto D & B procurement to improve integration. As a procurement route, depending on how it is set up, D & B permits supply chain integration more easily than the previous procurement routes discussed, and is placed in the middle of the project value chain.

Turnkey procurement has similarities to D & B, producing either a bespoke design for construction or adopting a standard offering by a turnkey contractor, who may subsequently customise it to an extent to suit client requirements. Turnkey contractors are normally involved early in the project value chain. Turnkey procurement offers greater integration of contractor-led design and construction with supply chain involvement. The contractual positioning and role of the designers will alter the impact on the project value chain. If the contractor employs the designers in house then there should be increased alignment in the project value chain. However, under turnkey, where the designer is independent of the contractor, another value system is imposed. The same argument applies to D & B. The earlier involvement of the turnkey contractor and greater opportunity for involvement of specialists within the supply chain in the design process place this above D & B in the schematic.

None of the procurement systems discussed above involves members of the construction industry supply chain beyond the commissioning and handover phase in the project value chain, i.e. into the use phase, although the members of the supply chain are supposed to consider the implications of the use phase during design development and construction. However, since the early to late 1990s, two new procurement routes have emerged which attempt to increase the degree of integration between the client and the construction supply chain from concept to use, notably the PFI and prime contracting. Whilst these procurement methods were not set up with this in mind, construction firms, owing to their increased knowledge of supply chain management and readiness to manage risk, have taken the lead in forming PFI consortia and prime contracting supply chains.

As indicated earlier, prime contracting was piloted through the BDB pilot initiative from 1997 to 2000. The initiative has three primary objectives:

1. To develop a new approach to construction procurement – prime contracting – based on supply chain integration.
2. To demonstrate the benefits of prime contracting in terms of improved value for the client and profitability for the supply chain through two pilot projects and a supporting 'tool kit'.
3. To assess the relevance of the new approach to the wider UK construction industry.

Prime contracting is now the primary procurement route of the MoD's Defence Estates organisation for capital works (capital prime) and maintenance works (regional prime). As originally envisaged, the prime contracting organisation was seen as bearing no resemblance to existing contracting organisations. The prime contractor is seen as any 'agency' capable of leading long-term supply chains and could be drawn from the ranks of any organisation with the capability of providing this type of leadership. The prime contractor role also brings with it the single-point responsibility of delivery from concept to use in the case of capital prime or single-point responsibility for operational delivery of a facility, including aspects of portfolio construction, in the case of regional prime. As a concept, it has permitted existing construction firms to take the opportunity of utilising their supply chain skills in such a role and has brought them into direct contact with procuring clients. Whilst it was originally conceived from within Defence Estates, certain private sector organisations have also used the capital prime approach to procure their facilities. Prime contracting uses current best practice in supply chain management, incentivisation of the supply chain, value management, value engineering, whole-life costing and performance, and risk management to deliver a facility. An output-based specification is produced by the client, as opposed to a technical specification, to permit, in theory, the prime contracting delivery team an opportunity to introduce innovation into the whole process. Prime contracting also acknowledges a two-stage briefing process. Through an integrated project team and a partnering philosophy, client representatives, end-users, the prime contractor, designers, constructors and specialist suppliers are jointly responsible for developing and delivering a project or operational maintenance service. The procurement route adopts the concept of supply clusters. Clusters are developed around key elements of a facility for delivery, involving client representatives, end-users, the prime contractor, designers, constructors and specialist suppliers, who are responsible for resolving design, construction and interface issues. 'Structure and frame' may be an example of an element around which a cluster is formed and led by a cluster leader. The prime contracting route also expects, under capital prime, that the prime contractor and supply chain retain responsibility for proving the whole-life costing model into the use phase. This is termed the 'proving' or 'compliance' period and lasts between three and five years. Normally, a facilities management organisation will operate the facility during this period. Prime contracting is placed above the other procurement routes discussed previously because it has been designed specially to bring together best practice in supply chain management and other related methods and techniques. There are similarities between prime contracting and the PFI, and further discussion of the former will be deferred until the PFI has been dealt with. Funding for the project comes from the public sector.

The PFI, introduced by the UK Conservative government in the early 1990s, and embraced with even more vigour by the current Labour administration, is a finance, design, build, operate and transfer (FDBOT) method of procuring and delivering physical assets for use as a service to public sector clients. As with prime contracting, the PFI process commences with a public sector client developing a statement of need in the form of an output specification. Normally, four or five PFI consortia, comprising funders, designers, constructors, facilities managers and other specialist supply chain members, pre-qualify to bid in successive stages for the PFI. The bid stages comprise the invitation to negotiate (ITN), best and final offer (BAFO) and preferred bidder – with the number of bidders reducing from four to two to one at each successive stage. At the ITN bidding stage, consortia develop their whole approach to the PFI, including design, construction strategies and whole-life cost models, and attempt to better a 'public sector comparator', namely the 'net present value' price if the public sector were to deliver the same project over the same period. This is the government's measure of value for money for awarding a PFI. A 'special purpose vehicle' (SPV) acts as a surrogate private sector client for the respective PFI once the contract has been let. The SPV will take responsibility for the whole-life management of the PFI facility, normally for a period of between 25 and 30 years. A PFI requires of the consortium a clear understanding of whole-life performance of physical assets from design through to operation, maintenance and renewal, including the relationship between capital cost expenditure and operational cost expenditure of a facility.

A PFI should, in theory, permit minimal disturbance to the project value chain, especially if the client has defined correctly the output specification and embedded its value system in its require-ments and descriptions. The PFI, unlike other procurement routes with a capital expenditure focus, also has a 'double-edged sword' since the chosen consortium is required to operate what it has designed and built for anything up to 25 or 30 years. There are simi-larities to prime contracting here. The focus of the consortium should be on continuity and integrity of the project value chain and product delivery throughout the PFI process. The PFI, before the introduction of prime contracting, was the only system where the 'value thread' could be maintained in the user value system. It also provides one of the greatest opportunities to leverage the principles of supply chain management, if the successful consortium chooses to adopt these. For this reason, it is placed at the top of Figure 12.4 for integration of the project value chain, with provisos in place that this is the intent of the consortium. Figure 12.5 sets out some of the issues concerning PFI procurement.

When one compares and contrasts prime contracting and the PFI as procurement routes, a number of similarities and differences come to the fore:

- Output specifications. Both procurement routes require cli-ents/end-users to produce specifications in output terms and not as a series of technical requirements. Output specifications

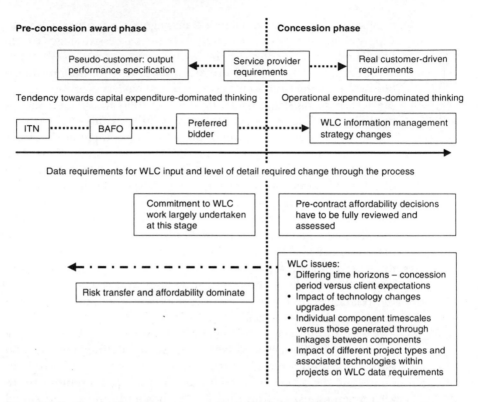

Figure 12.5. Pre- and post-contract award issues with the PFI (WLC, whole-life cost)

are essentially statements of functional requirements that permit different technical options to be developed provided they meet the required functions. This means clients must be able to state needs in output terms, a different way of thinking from traditional methods of producing technical specifications. There is a danger that these can become over-prescriptive and reduce the flexibility for service providers to innovate. As previously highlighted by Latham, these routes also raise the spectre of poor briefing (i.e. producing coherent and consistent output specifications), especially if clients are unable to easily articulate their needs explicitly and clearly.

- Both procurement routes require four or five consortia/supply chains to bid through three successive stages. Regardless of whether the route is PFI or prime contracting, they both go through similar bid processes. The PFI has been seen as a costly exercise to bid for, with sums of £2–3 million often quoted as the cost of bidding. Equally, regardless of whether the PFI or prime contracting is being considered, at the invitation to tender/ITN stage four or five teams could be developing different designs; at the BAFO stage this drops to two, with only one being chosen finally at the preferred bidder stage. The cost of failed bids will have to be recouped from somewhere, since a high proportion of the cost is at bidder risk. Equally, producing designs that are set aside by an unsuccessful tenderer could be argued to be wasteful of scarce industry resources.

- The PFI and prime contracting differ in their requirements for the use of best-practice supply chain management. It is a mandatory requirement as part of the prime contracting process; it was designed with this in mind. It is at the discretion of each PFI consortium as to how it structures the process from a supply chain perspective, both during the bid process and after contract award. The consortium may or may not reap the benefits of integrated supply teams, depending on the organisational and contractual structure.

- Defence Estates has produced a model form of contract for prime contracting, available on its website. A updated standard form for the PFI is to be published shortly by Butterworth. The bespoke nature of PFI contracts to date, and their level of detail, does raise issues about whether they will create their own climate of adversarialism when things go wrong, which at some stage has to be an inevitability.

- Whole-life management and performance. Both prime contracting and PFI require bidders to produce a whole-life cost model. Prime contracting, through its delivery structure, requires all key members of a supply chain to enter a dialogue to resolve capital versus operational expenditure early in the process. The PFI process, as noted above, ensures that the consortium, designers, constructors and facilities managers have to live with the consequences of their decisions for a protracted period of time. A PFI, provided it is structured to permit full supply chain dialogue to happen, can also resolve capital and operational expenditure issues early in the process. It will also test the adequacy of the life cycle cost model developed as part of the bid process. Owing to the funding regime of prime contracting, the supply chain consortium, led by the prime contractor, has a proving period of perhaps three years to test the life cycle model, particularly in terms of, for example, energy consumption. However, unlike the PFI, the proving period is of such a short duration in the context of the life of a facility that it is unlikely that the robustness of materials and components will be fully tested. This will only occur once the client/end-user has taken back the facility from the prime contractor.

- The public sector comparator adopted as a measure of value for money under the PFI has become the focus of much attention. An array of value management and value engineering studies conducted by the author has clearly suggested that designs have to be optimised in terms of capital and operational expenditure to produce, potentially, a good whole-life performance. Value management and value engineering provide a very structured process to make these decisions explicit. Equally, they have to be planned proactively into design development to ensure that this happens across the supply chain, including the client/ end-user. Unless this debate has happened during the process of developing the public sector comparator then its rigour as a true test of value for money is open to question.

- The discussion on the project value chain above has raised issues about the extent to which the client value system is able to be embedded into design development. The traditional approach was used as the benchmark, providing the maximum opportunity for the client value system to be embedded in designs. Questions were raised about the structuring of competitive D & B, since bid teams may be unable to have direct and sustained access to clients/end-users until after contract award. The same issue arises with the PFI and prime contracting. Both go through a three-stage bid process. During the invitation to tender/ITN stages, PFI and prime contracting teams may have very limited access to clients, in which case designs may have been developed without major or significant input that would embed thinking about the client's value system into design development. An important part of the project value chain is missing; the 'value thread' has to a significant extent been broken at important moments in design development. Given this, the process of developing output specifications becomes a key component of both the PFI and the prime contracting process; it is the only source of extensive commentary on the client's value system.

- Risk transfer. A primary objective of the PFI is to transfer risk to the private sector for an FDBOT project for 25 to 30 years, but at a price. The same applies to capital prime but to a lesser extent, since the supply chain consortium does not have to finance the facility and only has operational responsibility for the proving period. Under regional prime, the operational responsibility is extended up to ten years, but again the prime contractor does not have to finance the facility.

- Supply chain clustering. Prime contracting, through the BDB pilot initiative, developed the concept of supply chain clustering for project delivery. The concept has its origins in manufacturing and has been extended into both regional and capital prime. The formation of new, integrated supply teams does pose problems in getting to grips initially with the concept, especially where supply chain members have traditionally not been involved early on in the delivery process. They have to learn new skills. There are questions as to whether supply chain clusters are generic or specific. If they are generic across projects, under capital prime cross-project learning becomes much easier. If they are specific to a situation, then learning across projects becomes more difficult and mechanisms need to be in place to ensure this happens.

Since the focus of this chapter is on raising issues about innovative procurement methods when set against the backcloth of the project value chain, it is worth mentioning three other procurement methods that currently exist in the industry:

- The NHS Estates Procure 21 procurement method. This uses best practice in supply chain management, adopting principles that have stemmed from the BDB initiative for both capital projects and PFIs.
- The PPC 2000 partnering contract is one that has developed out of the Egan initiative. It purports to build in good practice on partnering, integrated supply chain management, value management, value engineering, risk management and whole-life performance. It has been widely piloted. The contract does not specify the use of supply chain clustering as a way of working. Partnering as a concept is about having an infrastructure of trust across the supply chain. At one level, trust is an interpersonal issue involving consistent behaviour. A partnering contract raises questions as to whether trust can be contractualised.
- Early contractor involvement. The Highways Agency has recently introduced a new form of procurement utilising the New Engineering Contract, where constructors are involved much earlier in the design development process to increase the level of supply chain integration. The approach does not specify the use of supply chain clustering as a way of working but sanctions the use of value management, value engineering, risk management and incentivisation, whole-life performance, and sustainability.

Summary

This chapter has focused on exploring a range of issues associated with new, innovative forms of procurement. The implications are clear. The agenda for change in procurement methods has been driven by major reviews of the industry. It will be driven in the future by those that have signed up to the Clients' Charter, and central government departments procuring construction services. These services will be procured using integrated teams and supply chains assembled on a long-term basis, and utilising supply chain skills at the right time in the delivery process. These initiatives will cascade into other parts of the public sector. This agenda sees construction aligned with manufacturing processes, where this supply chain integration is the norm.

Bibliography

Department of the Environment, Transport and the Regions. *The Egan Report: Rethinking Construction.* DETR, London, 1998.

Holti, R., Nicolini, D. and Smalley, M. *The Handbook of Supply Chain Management: The Essentials.* Publication C546. CIRIA/Tavistock Institute, London, 2000.

Kelly, J. R., MacPherson, S. and Male, S. P. *The Briefing Process: a Review and Critique.* Royal Institution of Chartered Surveyors, London, 1992.

Kelly, J. R., MacPherson, S. and Male, S. P. *Value Management – a Proposed Practice Manual for the Briefing Process.* Royal Institution of Chartered Surveyors, London, 1993.

Latham, M. *Constructing the Team: Final Report of the Government/Industry Review of Procurement and Contractual Arrangements in the UK Construction Industry.* HMSO, London, 1994.

Male, S. P. Building the business case. In: Kelly, J., Morledge, R. and Wilkinson, S. (eds), *Best Value in Construction.* Blackwell, Oxford, 2002.

Male, S. P. Supply chain management. In: Smith, N. J. (ed), *Engineering Project Management,* 2nd edition. Blackwell, Oxford, 2002.

Male, S. P. and Kelly, J. R. Value management as a strategic management tool. In: *Value and the Client. Proceedings of ICC Seminar,* pp. 33–34. Royal Institution of Chartered Surveyors, London, 1992.

Male, S. P., Kelly, J. R. Fernie, S., Gronqvist, M. and Bowles, G. *The Value Management Benchmark: a Good Practice Framework for Clients and Practitioners.* Thomas Telford, London, 1998.

National Audit Office. *Modernising Construction.* Stationery Office, London, 2001.

Standing, N. Value engineering and the contractor. PhD thesis, University of Leeds, Leeds, 2000.

Strategic Forum for Construction. *Accelerating Change.* Rethinking Construction, London, 2002.

CHAPTER THIRTEEN

Procurement through programme management

M. Graham and S. Male

Introduction

Programme management is a value-adding business function that interfaces strategic management and project management to provide sustained organisation-wide capabilities and benefits over time. It provides a structured framework or 'gestalt' supporting organisational learning and a company's core business by grouping, ranking, allocating resources to and coordinating projects as vehicles for change.

This chapter provides an overview of strategic management and the link with programme management. Subsequently, by exploring alternative definitions of programme management, the chapter locates its position within the overall organisational structure between strategic and project management. Interfaces between strategy, programmes and projects are explored with a particular emphasis on programme management. The factors involved in developing a programme are discussed with special prominence being accorded to planning and control, benefits, ranking, and value. The chapter concludes with a consolidated overview of programme management and its relationship to procurement strategies.

Strategic management, managing change and programme management

The intention here is only to provide a very brief overview of the concept of strategic management, concentrating on its interface with programme management.

The strategic management process answers questions about what the organisation ought to be doing and why, and where it should be going and why. Strategic management also involves making choices and managing change. An organisation's strategic decisions are likely to be long-term in nature and with distant horizons. Distant horizons equate to uncertainty and therefore the need for flexible or adaptable solutions. Other characteristics of strategic decisions include the need for integration between conflicting objectives and divisions within the organisation, and often the need for radical and possibly unpopular actions. Change, forming part of the strategic management process, is inherent in business organisations.

Change can be classified into two major and three contingent types:

- *Recurrent change* is incremental and routine, and requires no major realignment of the organisation with its external environment. *Operational change*, which occurs at the lower levels of an organisation through its day-to-day activities, is a form of recurrent change.
- *Transformational change* creates a fundamental shift between the organisation and its external environment. Transformational change can comprise strategic or competitive change. *Strategic change* is immediate, fundamental, radical and discontinuous. Most managers in an organisation are unlikely to see it coming. It affects the organisation from top to bottom. *Competitive change* creates a fundamental shift between the organisation and its environment in the medium to longer term. This will normally be felt as a sustained, deep-seated and continuous pressure on the firm to readjust its activities.

The role of an organisation's executive, for example the board of directors, trustees or governors, is to manage change and place the organisation at the optimum position within its environment. That will first require deciding where that position is. 'Placing' implies a forward position, action and change. Establishing where or what that forward position is requires a strategic decision. A strategy is a plan, a way of doing things, and as such strategy is pervasive. A plan could quite feasibly be to maintain a status quo, or it could be a vision of change. To be realistic and capable of achievement, the strategy must be matched to available resources, and so the plan or scope of activity is constrained within a boundary. Strategic management can therefore be described as 'defining a future position and matching resources to that vision'.

The process of creating a strategy is generally as follows:

1. Investigate the situation – define the decision to be made.
2. Develop alternative decisions – to ensure the right problem is being addressed.
3. Evaluate alternative decisions – options appraisal.
4. Select.
5. Implementation and follow-up.

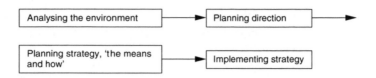

Figure 13.1. A simple model of the strategy creation process

This strategy creation process can be simply modelled as shown in Figure 13.1.

Items 1 and 2 above are normally termed 'strategy formulation'; items 3 and 4 are termed 'strategic choice'; and item 5 is termed 'strategic implementation'. It has been recognised for many years that implementation is frequently the necropolis of strategy. Strategy formulation tends to dominate over implementation.

Projects are an outcome of the strategic management process; they are experienced through strategic implementation. The increased use of projects has also brought the need to marshal project activity in some coherent and beneficial way. Programme management offers a structured, integrated and central approach to project selection and resource allocation so that the aims and objectives of the organisation as a whole can be balanced. Programme management is, therefore, a business function, providing information, not data, and is designed to support the core business of an organisation. As a business function, it must have demonstrable cost and time benefits. Programme management can also be viewed as providing the link between strategic and project management. It has developed in response to the widespread use of projects as a means of realising strategic change. As a link, its role is two-directional. First, it assists the strategic decision-making process, and second, it delivers the changes necessary. Its central position within the organisation enables it to ensure that the strategic delivery of each project is consistent with that of the other projects, and their own strategies. This means that the right projects must be selected, that appropriate scarce resources are allocated as and when necessary and that appropriate monitoring and control procedures are established.

Basics of programme management

Programme management provides an umbrella under which several projects can be coordinated. It does not replace project management, but rather it is a supplementary framework. Managed in this way, projects are more likely to be driven by business needs rather than personal agendas, the chance of duplication is reduced and interdependencies become more explicit and recognisable.

Part of the role of programme management is to contribute to that strategic decision-making process. By merging the boundaries between 'strategy', 'programme' and 'projects', it makes the relevance and importance of this interface, as the first step towards effecting change, become apparent, and this interface should not be undervalued. The need for this interface, as a communication conduit, is essential given the volatile, responsive nature of some strategic decisions (e.g. the reaction to an OPEC decision or major corporate bankruptcies) on the one hand, and the relatively long lead-in times of major projects and the complexity of their interdependencies on the other. Strategic decisions are by definition broadbrush, based on often incomplete or ambiguous information. The potential for changes to strategic requirements during the project development phase could be quite considerable. A permanent channel for communicating changed strategic requirements through to the implementation of projects is therefore essential.

It is argued here that it is not enough to just develop a strategy without considering its implementation. The interface between strategic management and programme management facilitates an opportunity for senior programme managers to advise on the feasibility of implementation. Equally, programme management may also influence the creation of strategy. By grouping together several small construction projects, for example, and adding in their facilities management requirements, a strategic decision to adopt a Private Finance Initiative-type procurement route could be made. In such a case it could be said that programme management had influenced, along this interface, a change in financial strategy. That example serves also to show another important face of programme management, which is its ability to add value or benefit to the organisation as a whole.

Although strategic management can be considered superordinate to programme management in that it provides a framework for the latter, it is not superior in the sense of being more important per se. It is rather that they have different roles to play in the same organisation and they complement one another. Having introduced programme management through its interface with strategic management, the next section examines in more detail the definitions, nature and scope of programme management.

Defining programme management, its nature and its scope

Programme management is not a new concept. It was probably first consciously used to deliver a defence initiative following the Second World War that became known as the Polaris missile programme.

Whilst network analysis and, largely, international peace have also been attributed to the Polaris programme, sadly the benefits of programme management have been largely overlooked since that time. Reasons might include the fact that the early developments in programme management were in the defence industries, areas that are often of a sensitive or secret nature and with functions that are essentially non-economic. Nevertheless, the need for an effective interface between strategy and projects has continued to exist, albeit largely unrecognised. However, with the current advent of several public sector initiatives and major infrastructure investments, the need for a distinct 'discipline' that can accommodate flexible management structures capable of responding efficiently to uncertainty and multiple goals has once more been recognised.

Project management is now a relatively mature discipline, having developed many excellent tools for the delivery of predefined objectives. It is concerned with the delivery of a unique piece of work. Programme management is extensively interactive with strategic management and individual projects and is instrumental in adding value to the organisation. It is the role of the organisation's strategic management team to identify and articulate the need for change. Projects are the instrument of change, a conduit through which the organisation's desires to improve, expand, adjust, etc. are delivered or converted into reality. The need for programme management becomes apparent when the scope of change is so extensive as to require delivery through several projects over protracted periods. The position of programme management in the organisation is between strategy formulation and delivery. Its ethos is a hybrid of strategic management and project management. It is a permanent, pivotal function in any changing or developing organisation, providing full-circle vision across the boundaries of strategy development and project delivery.

Unfortunately, there is no convenient dictionary definition of the word 'programme' for use within the context of business management. However, several of the limited number of writers on the subject of programme management provide insights into this emerging discipline. They have attempted to remedy that situation by contributing their own definitions or descriptions, which include:

- 'the process of co-ordinating the management, support and setting of priorities on individual projects, to deliver additional benefits to meet changing business needs ... a portfolio of projects which are managed in a co-ordinated way to deliver benefits which would not be possible were the projects managed independently' (Turner and Speiser, 1992)

- 'a grouping of projects, either for purposes of co-ordinated management or simply as a hypothetical construct to facilitate aggregate reporting at the strategic level' (Gray, 1997)
- 'the co-ordinated management of a portfolio of projects that change organisations to achieve benefits that are of strategic importance' (Office of Government Commerce, 1999)
- 'a collection of projects related to some extent to a common objective' (Association for Project Management, 2000)
- 'a framework for grouping projects and for focusing all the activities required to achieve a set of major benefits' (Pellegrinelli, 1997).

Whilst there has not yet emerged an unambiguous and universally accepted definition of programme management, this should not be seen in a negative light. In fact the opposite is more appropriate; unencumbered by definition, programme management is free to grow, innovate and develop to find is own level and identity.

It is tempting to consider large projects or multi-projects as programmes because there is an existing and extensive body of knowledge and experience covering projects and project management. Conversely, there is an absence of a coherent and widely recognised body of knowledge surrounding programmes and programme management. The difference between a 'programme' and a 'multi-project' can be illustrated by comparing the Polaris missile 'programme' and an Olympic Games stadium 'project' by looking at their respective time and strategic objectives/benefits aspects. Polaris was to be completed 'urgently'; there was no defined date. An Olympic stadium has to be completed by a fixed time. Polaris was designed to benefit the whole of the free world over a long period; a stadium's benefit can be said to be more parochial. A stadium could serve a limited population, and alternative arrangements are conceivable, albeit inconvenient. However, there was no option of transfer for a failed Polaris programme. A stadium's total funding arrangements are put in place before work starts. Polaris was funded by annual budget allocation.

The danger of attempting to employ project management techniques to the management of a programme is in the level of detail. This will destroy the inherent flexibility that programmes offer and, more importantly, ignore their involvement with the organisation's strategic management. Thus, programme management supports both strategy *and* projects. There are three parts to programme management:

1. Selecting projects.
2. Assigning priorities to projects.
3. Coordinating those projects by managing their interfaces.

The first two parts are clearly the more important and require a knowledge of and interaction with strategic management. A programme will provide a capability, i.e. a benefit, whereas a project (regardless of size or complexity) is only a process and enables a benefit to be obtained. A project has (or should have) a clearly defined objective which enables a benefit to flow. A programme will deliver a benefit. Programme management also provides a repeat-business/project framework, or way of thinking, that will absorb and retain the benefits of organisational learning. Amongst other things, that framework brings together related projects and maintains a strategic view over them, and aligns and coordinates them with a programme of business change.

Once the required projects have been identified, it will then be necessary to implement them. That is of course the role of project management, but under the overall control and direction of programme management.

Project ranking within a programme

Selection of the projects to be included in that programme is an exercise of major importance, and the process of benefits appraisal, ranking and prioritising achieves that aim, the subject of subsequent sections.

However identified, any range of beneficial and feasible projects will need to be scheduled into a rank order. This process should not be confused with prioritising, which is concerned with the timing of activities within a programme or the allocation of resources. Ranking involves placing the projects into a hierarchy, reflecting the effect they will have on or the extent to which they will contribute to delivery of the organisation's objectives, creating a new value activity or enhancing an existing value activity. If, for example, a car manufacturer wanted to introduce a new model and it had been decided that the optimum way to do so was via a purpose-built factory, that project would probably be placed higher in the hierarchy than re-roofing the existing factory. Clearly, without a clear strategic direction, it is impossible to establish a rank order. Without such direction, the usual result is to keep the options open or to do nothing, The rank order within the programme will be subject to change through circumstances beyond the programme manager's control. In such a circumstance it will be necessary to review and, possibly, realign priorities. Human resource opposition, discontent and demotivation will invariably accompany this action at least, which the programme manager must anticipate and attempt to cater for.

A project's rank may be considered its key to success. A highly ranked project will probably attract more resources, thus making its success more probable, provided it is feasible. It is also important to know how a ranking, be it high or low, is assigned, and often organisational politics will creep into this process. Unfortunately, many projects are considered to have failed because they do not achieve what was anticipated of them. Reasons for this could include:

- lack of clearly defined objectives
- unrealistic or over-optimistic objectives
- the project was driven by technology rather than business need.

There are four ways by which a project may secure preference:

1. *Differentiation* – establishing unchallengeable claims on valuable resources by distinguishing an organisation's own products from those of competitors, i.e. convincing others that only that particular project will satisfy the need.
2. *Co-optation* – attempting to absorb new elements into the decision-making structure as a means of averting threats to organisational stability or existence. This can involve either the claiming of power or simply sharing the burden of power.
3. *Moderation* – attempting to build long-term support by sacrificing short-term goals. The key to this strategy is an ability to estimate and compare short- and long-term gains and losses.
4. *Managerial innovation* – an attempt to achieve autonomy in the direction of a complex and risky project through the introduction of management techniques that appear to indicate unique management competence. This is often seen as high-level intervention, giving the impression of the project being important. However, this can often stifle innovation from lower down in the organisational hierarchy.

There is no formula that will uniquely identify those projects deserving a high ranking, but some formal approach will form the integrating factor in the array of choice. The evaluation framework must itself be a part of the strategic plan and be flexible enough to allow preselection of techniques and methodologies as appropriate. Also, programme management efficiency can be improved by improved procedures for project evaluation, i.e. a requirement for a rigorous and competitive analysis of the technical problems and benefits likely to be produced and their cost implications. The knowledge that there exists competition to secure a superior ranking will force competing project sponsors to examine their case and arguments more carefully. Finally, there is a likelihood that there may be many proposals to select from, and a screening process is required to eliminate early on any projects that are clearly deficient against preordained criteria, e.g. a minimum internal rate of return.

Generally speaking, the selection criteria should include the following questions. Does the project:

- take advantage of an organisational strength?
- avoid dependence on a known organisational weakness?
- offer an opportunity to gain competitive advantage?
- contribute to internal consistency?
- present an acceptable level of risk?

Finally, it should be noted that even if an individual project delivers according to its plan, it does not necessarily mean that the programme has added value.

The project-ranking process provides an excellent opportunity to implement the techniques of value management to give a programme or an individual project the best chance to add value. Value management is a structured, systematic, challenging, team-based process that keeps function and purpose to the fore and attempts to make explicit the package of whole-life benefits an organisation is seeking from a service, project, programme or product. It is not the intention of this chapter to discuss value management in detail; it has also been mentioned in other chapters.

The process of strategic management and its link with projects and programme management are demonstrated in Figure 13.2. Programme management has been defined in this chapter as an integrating business function to support the core business of an organisation. As a business function, it must have demonstrable cost, quality and time benefits and have developed in response to the widespread use of projects as a means of realising strategic change. To be effective, programme management requires the right projects to be managed, selected and coordinated and requires appropriate scarce resources to be allocated as and when necessary. It also requires appropriate monitoring and control procedures. It has already been stated that programme management's position in the organisation is between strategy formulation and delivery. As such, it has interfaces, distinct links or communication conduits between strategic management in one direction and project management in the opposite direction.

Programme management and procurement strategy

As indicated in Figure 13.2, programmes or a single project are an outcome of the strategic management process. They also contribute

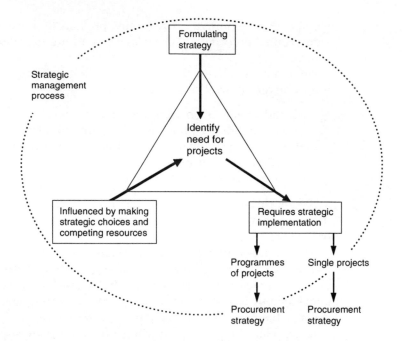

Figure 13.2. The link between strategy and programme management

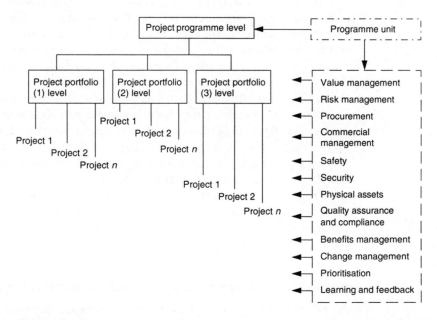

Figure 13.3. Programme-level procurement strategy

to it. Previous chapters have discussed in detail procurement strategies for single projects, and they will not be discussed further here in detail. However, the discussion will explore procurement options at the programme level.

Depending on the size and complexity of a programme, it is possible that it would be broken down into portfolios of projects with similar objectives, timescales, locations or complementary resources, as indicated in Figure 13.3. A programme unit would be set up to handle:

- programme strategy
- policy
- guidance
- coordination
- consistency across portfolios and projects
- other initiatives as set out in Figure 13.3.

The structure set out in Figure 13.3 permits long-term alliances to be formed either at programme level or at portfolio level. Alliancing would engender cross-project learning, continuous improvement and performance benchmarking.

Programme and portfolio structuring also enhances the capability to enter into partnering structures and collaborative forms of procurement. Clustering strategies (either generic or tailored) also come to the fore. Procurement strategies for a single project or across multi-projects within a portfolio can be adopted for integrating design, constructor and asset management teams, and also for maintaining a differentiated contract structure if so required. The important point is that value management provides the integrating structure for dealing with complexities across a programme, or within project portfolios, multi-projects or single projects. As a methodology, it can assimilate the structuring and simplifying of complexity at programme or portfolio level through tailored procurement strategies, by bringing the right team together at the right time.

Summary

This chapter has demonstrated that, whilst programme management is not a new concept, it is an extremely useful value-adding addition to any organisation contemplating change. The absence of an accepted definition of this management function is considered a powerful strength, allowing its future natural development. Programme management contributes to strategic and project management. Both of these managerial functions have been reviewed but, more importantly, the interfaces between them have been explored. It is by the effective management of those interfaces that considerable efficiency gains can be made. There is also scope for improvement provided by the umbrella approach of programme management. It provides an effective 'gestalt' for managing projects, in the form of resources for implementing change in a coordinated way to deliver business benefits. Programme management is a value-adding business function, responsible in part for the selection of projects, potentially using the application of value management tools to assist with this task and in thinking through customised procurement strategies.

Bibliography

Association for Project Management. *Body of Knowledge*, 4th edition. APM, High Wycombe, 2000.

Clark, P. Social technology and structure. In: Hillebrandt, P. M. and Cannon, J. (eds), *The Management of the Modern Construction Firm: Aspects of Theory*. Macmillan, London, 1989.

Ferns, D. C. Developments in programme management. *International Journal of Project Management*, 15(3) (1991), 148–156.

Gray, R. J. Alternative approaches to programme management. *International Journal of Project Management*, 15(1) (1997), 5–9.

Grundy, T. Strategy implementation and project management. *International Journal of Project Management*, 16(1) (1998), 43–50.

Johnson, G. and Scholes, K. *Exploring Corporate Strategy*. Prentice-Hall, Englewood Cliffs, NJ, 1989.

King, W. R. The role of projects in the implementation of business strategy. In: Cleland, D. I. and King, W. R. (eds), *Project Management Handbook*. Van Nostrand Reinhold, New York, 1988.

Langford, D. and Male, S. P. *Strategic Management in Construction*. Blackwell Science, Oxford, 2001.

Lansley, P., Quince, T. and Lea, E. *Flexibility and Efficiency in Construction Management*. Final Report. Building Industry Group, 1979

Office of Government Commerce. *Managing Successful Programmes*. HMSO, London, 1999.

Payne, J. H. and Turner, J. R. Company-wide project management: the planning and control of programmes of projects of different types. *International Journal of Project Management*, 17(1) (1999), 55–59.

Pellegrinelli, S. Programme management: organising project-based change. *International Journal of Project Management*, 15(3) (1997), 141–149.

Platje, A. and Seidel, H. Breakthrough in multiproject management: how to escape the vicious circle of planning and control. *International Journal of Project Management*, 11(4) (1993), 209–213.

Turner, J. R. *The Handbook of Project Management*. McGraw-Hill, New York, 1993.

Turner, J. R. and Speiser, A. Programme management and its information systems requirements. *International Journal of Project Management*, **10**(4) (1992), 196–206.

CHAPTER FOURTEEN

Future trends in construction procurement: procuring and managing demand and supply chains in construction

S. Male

Introduction

This chapter describes construction as comprising different demand and supply chain systems. It explores the impact of procurement routes upon these and highlights the fact that a demand and supply chain system in construction comprises two interrelated parts. First, there is the demand chain, which is client driven, and is set up by a chosen procurement route. This sets the parameters for demand chain management (DCM) in construction. Second, there is the supply chain, which is 'main contractor' driven, and responds to the demand chain of the client and the associated procurement route. Both 'chains' comprise the project value chain. By distinguishing between DCM (the strategic choice of a procurement route) and supply chain management (SCM), the main contractor's response to the demand chain, the chapter explores the potential impact of the recent changes in procurement routes that have emerged in construction as a result of industry restructuring. The chapter also introduces the concept of the 'strategic supply chain broker' and describes the role that the broker may play within the management of future demand and construction supply chain systems, depending upon the client type and chosen procurement route.

A paradigm shift has occurred in the procurement of suppliers, from being seen as an operational activity into one that is now strategic and linked to the long-term survival of a firm. SCM has emerged as a strategic function of a firm. Its purpose is to manage and integrate activities external and internal to the firm for the sourcing, acquisition and logistics of resources essential for generating products or services that add value to its customers. The manufacturing and retail industries have seen intense, prolonged, global competition. They have provided the main theoretical and practical developments in the field of SCM. The roots of the concept lie in physical distribution management, subsequently evolving into logistics management and then into SCM. Logistics management, with its origins in the military, is a planning framework concerned with optimising product and information flows within the organisation. SCM relies on trust and cooperation to work effectively and, unlike logistics management, its focus is on the internal and external integration of the supply chain and the management of materials, information and money through it. By addressing internal linkages within the firm and external relationships with suppliers and customers, the intent is to deliver superior customer value at less cost to the supply chain as a whole. SCM in construction is inherently linked to procurement strategies adopted by clients and their advisers and can add to or detract from adding value, depending on the chosen procurement method. SCM encompasses operational dimensions, and hence there are strategic as well as tactical dimensions to the concept.

The next section discusses a typology of demand and supply chain systems in construction, indicating their relationship to procurement methods.

A typology of demand and supply chain systems

Earlier chapters have discussed innovative forms of procurement and the move towards integrated teams. This section will reintroduce the client value system and the project value chain concepts, and emphasise a typology of demand and supply chain systems in construction, tailored to client types. This will lead into a discussion of the concept of the supply chain broker, how this new type of role may emerge as a response to the need for greater SCM in construction, and the alignment of this role with demand and supply chain responses depending on client types.

One of the important considerations for SCM in construction is the impact that the client (or customer) has on the process. The client commences the process of procurement, bringing together different skills through a procurement strategy to deliver a completed product – a physical asset of some type – to meet a need. All clients have distinct requirements and value systems, driven by their own organisational configurations, business and/or social needs, the external environment to which they have to respond and the manner in which they approach and interface with the construction industry. Earlier chapters highlighted the importance of maintaining intact the 'value thread' throughout the project value chain. Nigel Standing has segregated the project value chain into three distinct value systems:

- The client value system, concerned with stakeholder expectations, business drivers, needs and requirements.
- The multi-value system, bringing together different technical experts in design and construction from across the supply chain.
- The user value system, concerned with stakeholder expectations, business drivers, operational needs, requirements, fitness for purpose, and functionality.

Different procurement routes have the capability of integrating or segregating these different value systems. Figure 14.1 indicates the levels of complexity that can creep into the project value chain. The supply chain network in construction, as a set of contributors to the project value chain, can be classified as:

- Professional service firms, who provide a combination of skills and intellectual property to the process, typically comprising the designers and other professional consultants. Depending on the procurement route adopted, the delivery of professional services can fall within either the client's project-specific demand chain or the main contractor's supply chain.
- Construction and assembly firms. The on-site construction process is exceedingly skills focused. It comprises a range of different types of firms, which may have overall responsibility for the management of the process or for supplying inputs into the process. It involves fitting, installation, assembly, repair and on-site and off-site labouring. These firms are brought together at a location as part of the on-site manufacturing process.
- Materials and products firms. These comprise firms providing the materials, products and hired plant involved in the on-site process. These form part of the main contractor's supply chain and will involve the make (production), move (logistics) and store (stockpile) activities to and at a particular site location.

The diversity and structure of supply chain networks in construction is driven by the demand chains created by a diversity of clients. 'Main contractors' act as managers of a supply chain hub or node, and obtain their competitive advantage from three important supply chain activities. First, main contractors function as resource procurers and managers of supply chain networks on behalf of a diverse range of clients. They need to be capable of responding to

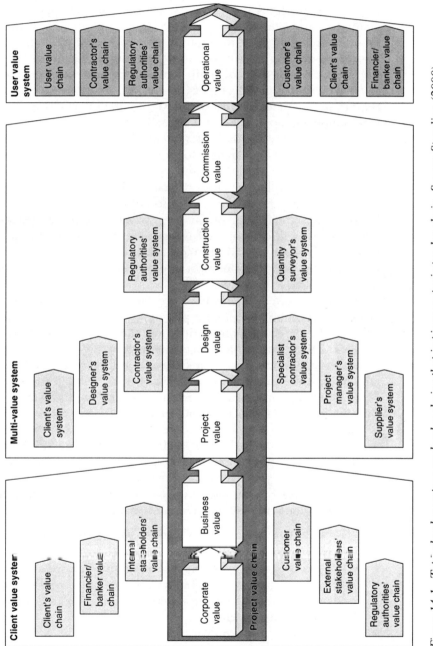

Figure 14.1. Typical value systems and value chains that impinge on project value chain. Source: Standing (2000)

the requirements of different demand chains, often created through a diversity of procurement and contract strategies. To respond successfully, they need an effective and efficient supply chain network. Second, they need to be good at logistics management. Third, they need to organise effectively and efficiently the on-site production activity at varied geographic locations, where the production activity of main contractors now involves extensive subcontracting. Effective SCM for the main contractor will require premeditated decisions on supply chain partners and strategic suppliers, those national/international firms from which they procure on a very regular basis or which are critical to project delivery. It is likely that tactical/operational supply chain decisions will be made from preferred and approved lists of suppliers for particular geographic locations. Equally, the 'hub' or 'node' capability of the main contractor needs to be developed more fully to incorporate standardised procedures, processes and components and an increased focus on delivering value to meet client requirements. As a comprehensive strategic activity, SCM will require main contractors to consider the most appropriate organisational structure to meet these demands.

In earlier chapters, clients have previously been characterised in terms of the economic demand placed on the industry in terms of volume – frequency and regularity – and the extent to which standardisation may exist from project to project in terms of parts, processes and design. Each type of project will be recapped briefly, and described subsequently as a distinct type of demand and supply chain system in construction requiring appropriate DCM methods through choice of procurement method. *Unique construction* is distinctive in terms of technical content, innovation, and the extent to which it pushes the barriers of the industry's skills and knowledge to the limit. With this type of project there is limited, if any, scope, for efficiencies in process or standardisation and repetition. This creates a *unique construction demand and supply chain system,* and typical DCM tools and techniques could include:

- use of competitive tendering coupled with strong pre-qualification and post-tender negotiation processes
- control over product delivery, exercised through specifications and forms of contract, and quality assurance processes for design and construction, including supervision and checking of workmanship during the latter
- a reliance on good professional advice.

Typical procurement routes that can work well under this system are the traditional route, construction management, management contracting in particular, and certain variants of design and build, all potentially overlaid with partnering structures to improve collaborative working.

Customised or *off-the-peg construction* tends towards the preceding, with the possibility for standardisation through repeat design, where clients may require only one or two similar physical assets, or where there is some foundation in previous designs undertaken for a range of clients but some adaptation is required for a new client. This creates a *customised construction demand and supply chain system*, and typical DCM tools and techniques could include:

- serial tendering with strong pre-qualification and post-tender negotiation processes
- control through appropriate forms of contract, use of performance specifications and intellectual property rights
- cross-industry collaboration.

The previous procurement routes can also operate well under this system.

Process construction occurs where there are repeat demands for projects and a high degree of standardisation is possible through the volume placed into the industry. Volume spends are highly probable. Efficiencies can occur from standardisation of design, components and processes. This creates a *process construction demand and supply chain system*, and typical DCM tools and techniques could include:

- use of forward planning and demand forecasting techniques
- rationalisation and consolidation of suppliers by spend
- use of strategic alliances, joint ventures and partnering with suppliers using non-contractual forms of agreement
- use of performance management, continuous improvement, quality circles, total quality management, just-in-time and inventory management, and lean supply systems.

Prime contracting, design–build–finance–operate, design and build variants, early-contractor-involvement procurement options and framework agreements can all work well under this system.

Portfolio construction occurs where clients have large and ongoing spends across a range of project types, with a diversity of needs in terms of technical requirements, degree of uniqueness, process or customisation, and content. Regular volume spends will permit long-term relationships with some suppliers. This creates a *portfolio construction demand and supply chain system*, and typical DCM tools and techniques could include:

- clustering of suppliers
- use of forward planning and demand techniques
- agile and flexible supply agreements, normally using some type of framework agreement or 'call-off' contract arrangements using schedules of rates and partnering philosophies
- use of the learning-organisation philosophy and supplier innovation, benchmarking and continuous improvement.

Portfolio construction, by its nature, could involve any combination of the above procurement systems.

This brief discussion has indicated the need to understand client types, their value systems, and that construction industry demand and SCM have to be attuned to the type of client. A schematic for the construction demand and supply chain system is set out in Figure 14.2. The next section explores contractor-led SCM in more detail.

Contractor-led SCM

The Warwick Manufacturing Group (WMG) research into SCM in construction identified the fact that first, the construction industry faces major problems with its suppliers that are fundamental to profitability. Second, world-class firms in other sectors of industry have developed structured, disciplined relationships with their suppliers to satisfy the needs of their final customers. WMG studied case examples from manufacturing, construction and the results from demonstrator projects as part of the research. Manufacturing companies involved in the research were drawn from the aerospace industry (3), oil operations (1) and shipbuilding (1). These case examples demonstrated some striking similarities to construction – small-batch production; a need to reduce cycle times; cyclic demand; complex products, often with high levels of customisation; and a need to drive down costs. However, there are also differences from construction, most notably the level of global competition faced by the firms in the case examples compared with those in a predominantly domestic construction industry. WMG concluded from its review of manufacturing that best-practice SCM in construction is not as advanced as that in the former, and that in the latter, client behaviour linked to the choice of procurement route is blamed for delays in implementing collaborative working methods. The research on construction also involved working with four construction firms and one major procuring client. Those firms in construction involved in the research that had implemented SCM were reported to have obtained increased work through negotiation by an amount of the order of 30–70% across the companies studied.

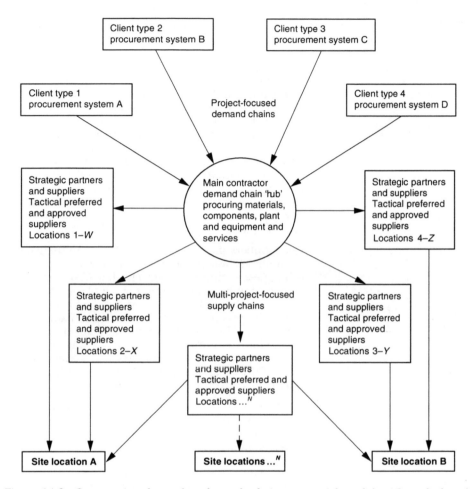

Figure 14.2. Construction demand and supply chain system. Adapted from Langford and Male (2001)

In addition, reductions in delivery time, increases in profit and market share, and advantages in competitive-tendering work can also accrue by bidding as a supply chain. As a consequence of its research with manufacturing and construction firms, WMG proposed a model for SCM in construction based on the 'Building Down Barriers' project. It comprises two principal components, namely organisational and project factors, each made up of separate activities. A schematic of the model is presented in Tables 14.1 and 14.2.

To summarise, construction has been argued to comprise a demand and supply chain system, the former driven by client requirements and the choice of procurement route. The latter is

Table 14.1. A model for SCM in construction: organisational factors. Adapted from Table A.1, Warwick Manufacturing Group (1999)

Organisational factors	SCM characteristics
Business development	There is a clear need to understand client business drivers, and this requires establishing long-term relationships with clients, where negotiation and not competitive tendering is the preferred mode of contractor selection. Contractors obtain a competitive advantage from their supply chain, and this can lead to business development opportunities for offering an integrated design and construct service involving suppliers, who are also client focused. The business development function will also need to address retaining teams to work on succeeding projects
Supplier sourcing	Key suppliers should be selected based on their skills, commitment to collaborative ways of working, a willingness to support contractor business objectives and continuous improvement. A supply chain champion should be appointed, preferably from within the procurement function of the main contractor. The main contractor and key suppliers should have a joint commitment to technology and process improvement, with a protocol setting out the rules of the relationship, which will be based on trust, openness, consistency, fairness and respect
Management of change	There needs to be a commitment from the main contractor's senior management to drive SCM into the organisation and to develop an appropriate strategy, including allocating resources to supply chain training. There needs to be a supplier measurement system implemented and a commitment to SCM established through pilot projects demonstrating measurable results and benefits

driven by the main contractor's response to the former. A typology of construction types and related demand and supply chain systems has also been proposed. The interaction between the project value chain, procurement and SCM has been developed further and a model of contractor-led SCM introduced.

The next section builds on the foregoing and explores the implications of the newer forms of procurement, involving collaborative working, different demand and supply chain systems, and SCM for the future operations of the construction industry. It argues that a new role in the industry will emerge – the strategic supply chain broker – as a natural extension of the drive towards collaborative

Table 14.2. A model for SCM in construction: project factors. Adapted from Table A.1, Warwick Manufacturing Group (1999)

Project factors	SCM characteristics
Management of the design process	Client needs must be to the forefront. Functional specifications should be adopted, using a structured and formalised design process to optimise functionality and minimise cost, and employing value management and value engineering. An integrated approach to design should be adopted, with the main contractor offering single-point responsibility for project delivery. A clustering strategy should be adopted for design development using suppliers and users from the outset. There should be a commitment to risk management and risks allocated to those best able to manage them. Information technology should be utilised to improve the communication of design, cost and planning information
Cost management	Profit and overheads should be agreed up front, with target cost and incentive mechanisms utilised to drive improvements. A formal, documented system of value analysis should be adopted throughout, with costs readily understood and transparent to all
Management of the construction process	Best practice should be adopted throughout and documented. Planning for the construction stage should commence during detail design and should involve suppliers within cluster groups. Continuous-improvement teams should be utilised extensively to remove waste, and resources should be allocated to team training. If the preceding is implemented then, consequently, quality checks on suppliers become redundant

methods of working, with supply chain competing against supply chain as a matter of course in the future.

The procurement, operation and management of future construction demand and supply chain systems

A programme of work was conducted during the period 1994–2000 that postulated possible future scenarios for the structure of the construction industry around demand and SCM. The European Large Scale Engineering Wide Integration Support Effort, eLSEwise for short, was concerned with contributing to the continued success of the European construction industry by assisting the different sectors and parties in their efforts to improve competitiveness. Large-scale engineering (LSE) projects are complex, multi-disciplinary, engineered undertakings involving design, construction and potentially operation, and are encountered at the top end of construction activities. Typical examples include complex buildings, process plant, infrastructure and other significant civil engineering works. An LSE project has many of the following attributes:

- High capital cost.
- Long duration and programme urgency.
- Technologically and logistically demanding.
- The requirement for multi-disciplinary inputs from many organisations.
- The creation of a 'virtual enterprise' for the execution of the project, that is, a group of organisations collaborating as a 'partnership'. The 'virtual enterprise partnership' could have longevity, and could design, construct, operate and maintain the end product until decommissioning, and could execute other LSE projects. It could be, however, a transient virtual partnership, which disbands after the project is executed and handed over or after decommissioning.

eLSEwise concluded that there would be a major shift of responsibilities from client to contractor, with the latter expected to take more risk. In addition, LSE clients would be outsourcing non-core activity and decreasing in size, with a consequent loss of in-house construction expertise. The research also confirmed a major shift in procurement routes over the next decade, with increasing use of strategic alliances and integrated supply chains in the LSE sector. Profitability, capital cost, whole-life costs, health and safety, and the timescale of projects were identified as key value drivers, and there was an increasing focus on value for money from clients. Greater use of information and communications technology (ICT) to assist project processes was also seen as an important driver. The eLSEwise research also focused on the competitive advantage of LSE contractors, with key sources of competitive advantage including:

- the capability to provide attractive financial packages
- the ability to build winning alliances
- the ability to accept and manage risk
- the ability to invest in sales and in research and development
- the identification of client/user needs through market research
- the ability to procure on a global basis
- technical expertise and the right technology
- the integration of local and global knowledge
- political backing.

The eLSEwise consortium also identified that an LSE contractor's core competencies were widening. The LSE contractor of the future would be a project-centred organisation able to provide flexible logistics skills, manage human resources, provide technical construction skills, organise a network of specialists, have the ability to organise and control financial packages, and manage a complex multi-layered and multi-skilled organisation. In combination, these competencies would deliver an integrated offer.

The ideas raised in this work were extended further to include investigating the consequences of the above for the domestic UK construction industry using the structural-steelwork supply chain as an exemplar. The 'optimum solutions' research (Brown *et al.*, 2000) proposes that construction supply chains in the UK will be much wider, operating regularly with different support structures and also with a greater reliance on ICT. On a more regular basis, construction supply chains will include funders, designers and other members of the design team; a 'general' contractor; specialist contractors and suppliers of components and providers of services; and facilities managers. Their offering will be a total service package to the client. Using current trends in the industry stemming from the increased use of design and build, examples such as Mace's 'Branded Product' and, more importantly, the Private Finance Initiative (PFI) and prime contracting, the 'optimum solutions' research hypothesised that new roles will emerge in the industry, namely different types of *strategic supply chain brokers*. It was also hypothesised that there will be a requirement for *phased, seamless teams* to emerge and become the norm.

Historically, management contracting, construction management, and design and build are attempts at drawing together design and construction interfaces and responsibilities. These other mechanisms to overcome contractual adversarialism, and the need for single-point contact and responsibility are important drivers for the possible emergence of the strategic supply chain broker. The important facet of the role is that the broker is able and willing to take away risk from the client as a one-stop shop supplier of skills, expertise, product components and knowledge, at a price. Brokers will compete on their expertise using their supply chain base.

A typology of brokers and their skills base

Whilst the concept of the broker has similarities to that envisaged by Holti *et al.* for prime contractors in the Building Down Barriers project, it goes further. It extends the work and draws on prime contracting and the PFI. The role postulated here is for an organisation to emerge that will broker skills and finance, take and manage risk, and deliver asset value throughout the project value chain, into the end product and potentially beyond. At the upper echelons of broker firms, 'world-class' organisations would emerge, capable of total systems integration, to compete nationally and internationally, potentially at project and programme level.

This section proposes that a range of broker types will emerge. First, brokers could decide only to respond to pre-demanded products, the normal situation in construction. Second, brokers could go one stage further, not only offering the preceding service but also undertaking market research to understand client needs and requirements and then deciding to create demand on behalf of the clients by anticipating their needs in advance. This type of broker would use specialist staff experienced in tracking trends in clients' business environments and understanding their strategic and operational requirements. It is likely that such brokers would specialise in particular client types and projects. They would approach them with proposals to deliver and/or manage the types of physical asset they need. Third, two other forms of broker could emerge, based on type of demand but also, more importantly, on the requirements of knowledgeable and less knowledgeable clients. In this instance, using the Pareto principle, approximately 80% of all projects could be categorised as 'routine', comprising the 'customised' and process types of construction. This would act as one of the drivers for working with the same project teams to develop cross-project learning and to develop standardised processes, systems and procedures. This would form the basis of volume delivery within the industry, and the emergence of 'volume brokers' developing a brand reputation for timely, regular delivery to cost at an appropriate quality and functionality on relatively straightforward projects. Some brokers would emerge to deal with unique construction, where approximately 20% of projects could be categorised as innovative and leading-edge. The current management forms of procurement could be the drivers behind the emergence of this type of broker. These brokers would develop brand reputations as 'innovative brokers', with the requirement to deliver innovative solutions. Brokers operating under these conditions would have a pool of leading-edge supply chain members that would be niche providers to the industry. Finally, brokers could emerge that provide a one-stop shop for design and construction only, similar to design and build but working consistently with integrated teams; or brokers could emerge that would provide the total package from cradle to grave – concept, design, construction and asset management. These brokers would emerge from within the ranks of firms currently delivering prime and PFI contracts.

Table 14.3 draws the above together to propose a typology of brokers. Broker types 3 to 8 will require a high level of knowledge

Table 14.3. Types of strategic supply chain broker

Type of project	Type of demand			
	Pre-demand-led service only using integrated, seamless team	Combined service for pre-demand-led and market-forecast-led service	Total package: combined pre-demand-led and market-forecast-led cradle-to-grave service	Total package: combined pre-demand-led and market-forecast-led cradle-to-grave service
Routine, volume market	Type 1: design and make to order	Type 3: forecast, design and make to order	Type 5: forecast, design, make to order and operate	Type 7: forecast, design, make to order, operate and manage at programme level
Unique, niche market	Type 2: design and make to order	Type 4: forecast, design and make to order	Type 6: forecast, design, make to order and operate	Type 8: forecast, design, make to order, operate and manage at programme level

and skills that go beyond construction. It is also likely that as one moves from type 3 to type 8, the additional expertise of supplying finance as part of a package is likely to increase, depending on whether the client is in the public or private sector.

Given the range of expertise required to deliver the service, broker skills would require a 'large company' knowledge base, with existing long-term relationships in place with key members of the supply chain, and founded on trust. The 'optimum solutions' project postulated that brokers could emerge from a number of quarters, namely the ranks of the major national or international UK contractors, construction managers, consultant project managers, management consultants or international consulting engineering firms. Banks with strong interests in construction activity, perhaps funding build–operate–transfer or PFI projects internationally, could also act in a broker role using a specialist division acting as the interface between the bank and the industry. In addition, regional broker firms could emerge to cater for the smaller, occasional procuring clients. However, it is likely that if the broker types emerge as postulated above, construction would consolidate into a very hierarchically structured industry, with broker firms acting as the principal point of contact with clients. Should this arise, construction would restructure in the UK and concentrate around a major grouping of brokers, some competing nationally, some competing globally. Other parts of the UK construction industry would comprise part of each broker's supply chain system, at the local, regional, national or international level.

The broker's authority, as supply chain leader, would be derived from:

- previous experience and a core competence in market knowledge
- SCM expertise
- in-depth knowledge of the construction process but not necessarily on-site delivery skills, which could be subcontracted in
- a capability to structure and manage the total delivery process
- an ability to package skills and expertise, take and manage risk, and deliver and manage the project value chain in order to meet customer requirements using supply chain members.

Brokers would compete on the basis of their supply chain networks, with 'strategic team' capabilities derived from supply chain partners and strategic suppliers, and with tactical capabilities derived from a team composition using preferred and approved suppliers attuned to client, project type and geographic location. Depending on the service offered, brokers would include asset management within their portfolio of skills and services, with supply chain member choice or corporate configuration reflecting this alternative. It is also postulated that over time, brokers and their supply chain delivery teams would develop 'brand types' and have a reputation for delivering facilities of a particular functionality, aesthetic type and quality. The broker, as a strategic network coordinator, would determine the scope of work for teams, agree appropriate profit levels and balance differing team expectations, and would expect risks and value-based, rather than cost-based, rewards to be shared. Supply chain networks would operate under systems of open book accounting. The broker's responsibility would be to meet customers' needs, guarantee certainty of delivery and achieve client satisfaction.

Having identified eight different broker types competing on the basis of stable supply chains, we shall review in the next section the concept of the phased, seamless team as a complementary concept.

The phased, seamless team
Traditional, contractually based relationships can stray into adversarial interactions when problems occur. The Latham and Egan reports attest to this, as has every major review of construction since World War 2. Traditional procurement is also more likely to be linear and sequential, with one task following another and with different organisations procured at a particular point in time to suit

a particular project programme. This may militate against securing the right skills at the right time regardless of position within the supply chain. The 'optimum solutions' study proposed that underpinning the concept of the strategic supply chain broker is the use of a phased, seamless team. Seamless teams would pre-exist around a particular broker. Teams would be familiar with each others' working practices owing to long-term relationships and would recognise the value of timely information exchange to manage key project interfaces throughout the supply chain. The seamless team would operate on the basis of a moral and psychological contract founded on trust. Process and procedure would not be contractually based, and rewards and incentives would be based on performance indicators tied into client-focused value and not cost-driven service delivery. Teams would operate in a solution-oriented culture where pooling and sharing of information would occur using a combination of face-to-face problem solving supported by information technology. Project processes would dictate when team members came in and others left as the project progressed, and would require the team to operate with seamless handover processes and procedures. Team expectations, roles, responsibilities, skills and competencies would be fully understood among team members. Under this approach, deeper levels of commitment and cross-project learning would occur. The principles behind the seamless team would permit concurrency of inputs and skills for the benefit of the project and product, buildability, and decision-making taking full account of all aspects of the total process. Interfaces would be owned and information flows improved since they would be internal to the team. This method of working would lead to savings in time and money, and greater opportunities for continuity of work and learning.

The next section draws the chapter together to reach a series of conclusions about future demand and supply chain systems in construction.

The future

The construction industry comprises numerous project-based demand chains that are created through the procurement process for individual clients. The choice of procurement route is a strategic decision made by the client and/or its advisers and has a fundamental impact on the project-based demand chain. It has the capacity to assist or hinder the transfer of value through the project process – the 'value thread'. Main contractors occupy a 'node' or 'hub' role in the demand and supply chain system. Their role is to balance the competing needs of different project-based demand chains with the procurement of their own multi-project supply networks. Main contractors, through their supply networks, are ideally placed to become supply chain leaders working directly for clients. However, as the analysis of the broker role has shown, this may not necessarily be the case in the future.

Equally, contractors as 'manufacturers and assemblers' face considerable end-product diversity due to different client types, their individual requirements and design team influences, coupled with the impact of the choice of procurement route on roles and responsibilities. Regular, knowledgeable, volume-procuring clients are in a position of considerable market power to influence their own project-focused demand chain. However, whilst they may place considerable volumes of business into the industry, they are not in the majority numerically. Infrequent procurers, numerically greater in number, are in a much less powerful position to influence their supply chains, owing to much lower-volume, *ad hoc* spends. Equally, some contractors have the advantage of organisational size and can influence their own multi-project supply chains owing to their own market purchasing power; others have much less capability in this area. The author's experience from studies of pathfinder prime contracts, PFI and partnering also indicates that agreeing incentive systems through the supply chain is one of the more difficult areas. It depends on the procurement system and contract chosen.

One of the most comprehensive examples of implementing the early-involvement methods described above for demand and SCM in construction currently is prime contracting. The Highways Agency is also experimenting with early-involvement methods, and the PPC 2000 contract for partnering also permits this. The PFI, like prime contracting, is output and not input specification based. It has been applied in a variety of ways, including application to customised, process and portfolio client asset bases. A consortium supplying a finance, design, build, operate and transfer asset under a PFI, perhaps for up to 25 years, has an opportunity of delivering improved value for money throughout its supply chain when optimising capital and through-life costs. However, in general, PFI does not have a mandatory requirement to use structured value management, value engineering, and risk management, partnering, SCM, clustering or incentivisation procedures, although good practice would suggest that these should be adopted. NHS Procure 21 integrates these aspects formally. Under the PFI, consortia have to live with the consequences of the decisions made early in the project life cycle for a long time. Prime contracting and the PFI become powerful systems for procuring volume asset bases if both involve the mandatory requirements set out above. Both would require trust to be widespread within the industrial supply chain base of construction. Evidence from research studies conducted by the author does, however, suggest that within some PFI consortia the use of existing procurement routes is being reinforced under the guidance of the 'special purpose company', without the movement to greater integration within the supply chain that the system could provide.

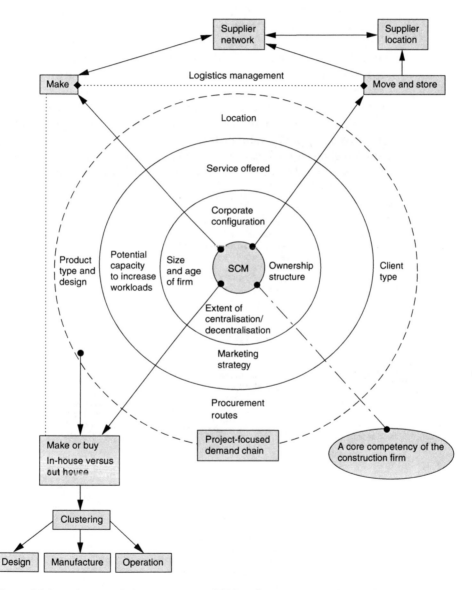

Figure 14.3. A framework for construction SCM in the construction firm. Adapted from Male (2002)

Figure 14.3 draws the preceding together and sets out a framework for thinking about SCM within the construction firm.

This chapter has also postulated the emergence of a new role in construction, the strategic supply chain broker, as a response to increasing levels of collaborative working with supply chains. The PFI and prime contracting are seen as the most likely procurement routes to lead to the emergence of a strategic supply chain broker in the industry. Brokers, emerging as supply chain leaders and coordinators, would be single-point contacts for clients, brokering skills and finance, working with long-term integrated, phased, seamless teams of supply chain members capable of offering capital investment or cradle-to-grave solutions. Strategic teams would comprise supply chain partners and strategic suppliers, with brokers deciding on the composition of tactical teams, drawn from preferred or approved suppliers, depending on client or project requirements and on geographic location. The exploration of the broker concept has also included differentiating between brokers that may work

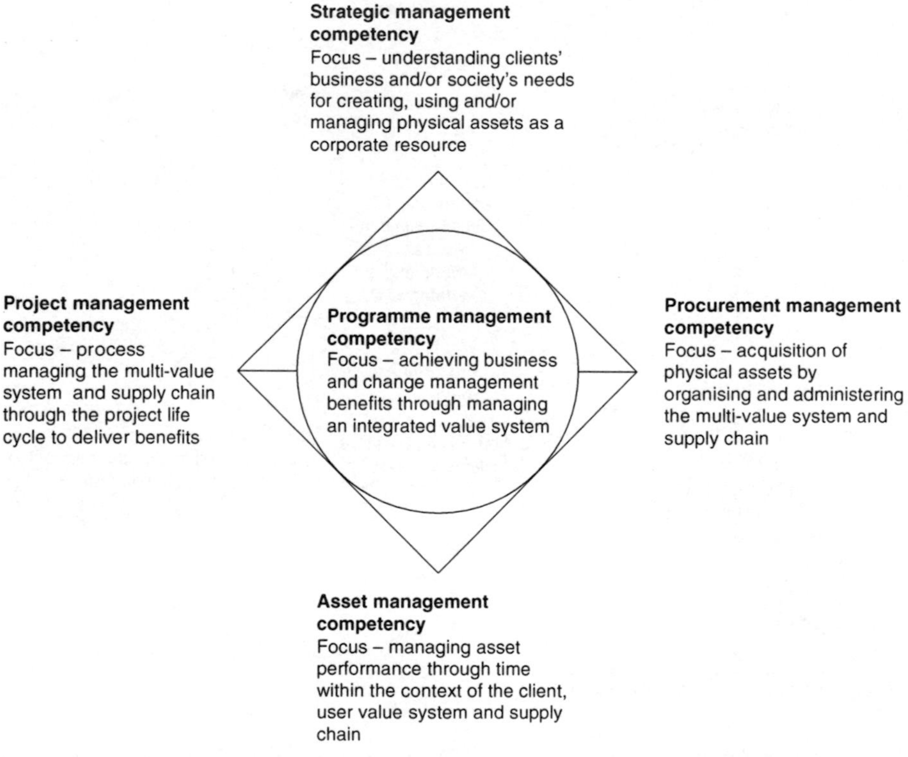

Strategic management competency
Focus – understanding clients' business and/or society's needs for creating, using and/or managing physical assets as a corporate resource

Project management competency
Focus – process managing the multi-value system and supply chain through the project life cycle to deliver benefits

Programme management competency
Focus – achieving business and change management benefits through managing an integrated value system

Procurement management competency
Focus – acquisition of physical assets by organising and administering the multi-value system and supply chain

Asset management competency
Focus – managing asset performance through time within the context of the client, user value system and supply chain

Figure 14.4. A framework for broker competencies

Table 14.4. Clients, demand and supply chain systems and the emergent broker role[a]

Response from the industry	Private sector clients						Public sector clients			
	Knowledgeable – regular procurers			Less knowledgeable – infrequent procurers			Knowledgeable – regular procurers		Less knowledgeable – infrequent procurers	
	Consumer clients: large owner/occupiers	Consumer clients: small owner/occupier	Speculative developers	Consumer clients: large owner/occupier	Consumer clients: small owner/occupier	Speculative developers	Consumer clients: large owner/occupier	Consumer clients: small owner/occupier	Consumer clients: large owner/occupier	Consumer clients: small owner/occupier
Unique				✓ Types[a] 2, 4, 6					NA	NA
Customised		✓ Types 1 and 2	✓ Types 3 and 5 or 4 and 6	✓ Types 1–6	✓ Types 1 and 2			✓ Types 5–8 + finance	NA	NA
Process	✓ Types 1, 3, 5 and 7	✓ Types 1 and 3					✓ Types 5–8 + finance	✓ Types 5–8 + finance	NA	NA
Portfolio	✓ Types 1–8		✓ Types 1–6				✓ Types 5–8 + finance		NA	NA
DCM orientation	Sophisticated leaders Internal advisers	Followers External advisers	Sophisticated leaders Internal advisers	Reluctant followers External advisers Wait and see	Reluctant followers External advisers Wait and see		Sophisticated leaders Internal advisers	Sophisticated followers Internal and external advisers		

[a] 'Type' refers to type of broker; NA, not applicable

purely in the context of single-project delivery and those that may work also at the project programme level. Figure 14.4 draws together the skill base that would be required by the different types of brokers. It also reflects key themes surrounding procurement.

Table 14.4 indicates that different broker types will gravitate to certain types of clients. It acknowledges that certain broker types would cater for the needs of the smaller, irregularly procuring clients. The problem remains, however, one of the cost of collaborative approaches. It is likely that in such instances, the smaller, irregular procurers will continue to adopt more traditional, non-collaborative approaches to construction. If this is the case a structure for the industry different from that of today will emerge, with large corporate clients served by leading-edge supply chains, working collaboratively, and small, irregularly procuring clients functioning under traditional 'adversarial' approaches.

Summary

The arguments set out in this chapter combine theory with extrapolation, with argument and counter-argument about the pluses and minuses of SCM for procurement. The debate has been supported by an investigation of the major forces that are now shaping and changing the nature and structure of the industry. The chapter has also been intended to offer the reader a series of models that can be used in practice to assist them to think through the issues surrounding procurement in construction in the future.

Bibliography

Brown, D. Williams, P., Gordon, R. and Male, S. P. *Optimum Solutions for Multi-storey Steel Buildings.* Final report. Department of the Environment, Transport and the Regions, London, 2000.

Centre for Strategic Studies. *Construction 2000.* University of Reading, Reading, 2000.

Croner. CCH. Management of Construction Projects, pp. 2-541–2-543. Croner. CCH, London, 1999.

Hassan, T. M., McCaffer, R. and Thorpe, T. Emerging client needs for large scale engineering projects. *Engineering, Construction and Architectural Engineering Management,* **6**(1) (1999), 21–29.

Kelly, J. R. and Male, S. P. *Value Management in Design and Construction: the Economic Management of Projects.* Spon, London, 1993.

Male, S. P. and Mitrovic, D. Trends in world markets and the LSE industry. *Engineering, Construction and Architectural Engineering Management,* **6**(1) 1999, 7–20.

Male, S., Kelly, J., Fernie, S., Gronqvist, M. and Bowles, G. *The Value Management Benchmark: a Good Practice Framework for Clients and Practitioners. Published Report for the EPSRC IMI Contract.* Thomas Telford, London, 1998.

Male, S., Kelly, J., Fernie, S., Gronqvist, M. and Bowles, G. *The Value Management Benchmark: Research Results of an International Benchmarking Study. Published Report for the EPSRC IMI Contract.* Thomas Telford, London, 1998.

Mitrovic, D. ESPRIT 20876, eLSEwise. In: *Swedish National Workshop.* Deliverable D103.

Mitrovic, D. and Male, S. P. *Pressure for Change in the Construction Steelwork Industry – Solutions and Future Scenarios.* Publication 293. Steel Construction Institute, Ascot, 1999.

Warwick Manufacturing Group. *Implementing Supply Chain Management in Construction.* Project Progress Report 1. Department of the Environment, Transport and the Regions, London, 1999.

Index

Page numbers in *italics* refer to tables and figures.